About *Commercial Fiction*

"In *Commercial Fiction* you experience everything a best-seller should be. The shakers and movers of business, government and lithe bodies collide with the sitters and squirmers of real life, as the powerful, the beautiful and the enticing exploit the average and the obsequious in their relentless pursuit of wealth, power and glamour."

— *Times-End Literary Supplement*

"Vernon Frazer rivals Joseph Heller and Nathanael West as a master of literary cartooning. *Commercial Fiction*'s living caricatures pursue their ambitions through a comic maze of game shows, talk shows, news shows, prime time features and late-night sci-fi reruns, pausing to endorse products at every station break. *Commercial Fiction* is a wild, wacky and wonderful roller coaster ride through the amusement park of contemporary television..."

— Los Angeles *Noon-Times*

"...a madcap world in which commercials interrupt reality and reality interrupts commercials faster than you can surf with your remote. Place *Commercial Fiction* at the top of your reading list!"

—*Krakus Reviews*

"Not since Chauncey Gardiner in *Being There* has a character risen to world prominence as meteorically as Ralph Putz in *Commercial Fiction*....A laugh-a-minute literary *tour de force*..."

—Manhattan *Morning Star*

"The most hilarious novel I've read since *Catch-22!* A must read!"

—Chicago *Evening Messenger*

"A funny, fast-paced read, *Commercial Fiction* is literary video-surfing at its satirical best. Highly recommended."

—Seattle *Sun-Dispatch*

"Is it hyper-reality, or is it *Commercial Fiction*? Only the author knows for sure...Or does he?"

— San Francisco *Midnight Star*

"Franz Kafka scripts *Saturday Night Live...Commercial Fiction* portrays a media-saturated world that will make its readers question their ability to tell fantasy from reality."

—Detroit *Auto-Examiner*

"Delightful, hilarious and incisive...A bold and biting commentary on contemporary life. Don't miss this one."

—*Publishers Millennial*

Commercial Fiction contextualizes the lives of the shakers and movers of business, government and lithe bodies in the most contemporary of literary environments: the Post Nomo-pomo aesthetic of interactive fiction that embraces our society's demand for Instant Product while satirizing it at the same time."

—*Quarterly Journal of* Commercial Fiction

Commercial Fiction

Also by Vernon Frazer

POETRY

Improvisations (XXV-L)
Amplitudes
Improvisations (I-XXIV)
Demolition Fedora
Free Fall
Sing Me One Song of Evolution
Demon Dance
A Slick Set of Wheels

FICTION

Relic's Reunions
Stay Tuned to This Channel

RECORDINGS

Song of Baobab
Slam!
Sex Queen of the Berlin Turnpike

ANTHOLOGIES

THE JAZZ VOICE
THOMAS CHAPIN–ALIVE

Commercial Fiction

by

Vernon Frazer

Beneath the Underground

ACKNOWLEDGMENTS

An earlier version of *Commercial Fiction* appeared in serial form in *Plain Brown Wrapper.*

Cover Design: joey Zone

ISBN 0-9633465-9-8

Second Printing

Beneath the Underground
568 Brittany L
Delray Beach, FL 33446

Printed in the United States of America

To Dick Freeman,
a friend who's published my most adventurous work

COMMERCIAL FICTION

is brought to you by

THE AUTHOR*

(Medium shot. A SUBURBAN MOTHER *and her* SON *stare at the present page, hovering on display above a table in a Megabook$ chain store.)*

Son: Mommy, what's that?
Mother: That's the author.
Author: *(From behind the page.)* That's right. An author, ninety-nine and 44/100 percent pure. His work is guaranteed to build strong minds in eight ways, or your money back.
Son: I want to see him. *(Steps forward, tears away the page, revealing the author's snarling, bearded face.)*
Author: You little brat!
Mother: What a rude man! We're not buying *his* book. *(Drags* SON *by the arm.)*
Author: *(Runs after them.)* I'm *sorry*. I'll even *give* you a copy. Here!
Mother: I wouldn't even *take* one from the likes of *you.*

(The AUTHOR *steps toward the camera, pawing his gray beard into a strained expression of congeniality.)*

Being an author isn't as easy as it seems. It looks glamorous on the surface: the instant wealth, the Celebrity Status, the hundreds of adoring women

---

* an unco-opted non-subsidiary of WalMart, BrandsMart, K-Mart, Telemart, AT&T or any other conglomerate in the American Merchandising Machine (as of this writing).

just waiting in line for an autograph, a few kind words and maybe a little *tete-a-tete* afterward.

(The book table behind the AUTHOR *has no visitors.)*

Before you can become an author, you have to be a writer. Being a writer takes lots of hard work and dedication. If your dedications are to editors and publishers, being a writer doesn't take quite as much hard work.

But there's more to being a writer than the act of writing itself. Writing requires courage, the kind of courage that enables you to:

1) Suck up to editors, publishers, agents and semi-literate staff in chain bookstores.

2) State your heartfelt personal opinions so that they agree unequivocally with those of the people whose influence can make or break your career.

3) Give your body and soul to people who wouldn't even consider buying them at a clearance sale.

Last but not quite least, there's the act of writing itself. Writing demands time, patience and a head thick enough not to crack while you're pounding it against a brick wall for decades on end. If you're an especially good writer, you'll need stamina so that you can continue to produce work posthumously. That's when the demand for your work gets really strong. To develop the stamina you need, you have to work hard, live clean and die young. That way, aging doesn't deplete the bodily energies you need to continue producing posthumous work. If you'll give me just a few minutes of your time, I'll be more than happy to show you what it's really like to live

A Life in the Day of (1)

I heard a fly buzz when I lived.

6:44 a.m. zzzzzzzzzzzzzzzzzzzzzzzzzzzzzzzzzzzzzz

spooned around her Florida motel neon flickering outside window bodies naked on an empty beach sunset moon felt her softness what a way to

6:45 a.m. ***BZZZ-BZZZ-BZZZ-BZZZ—BZZZ-BZZZ-BZZZ-BZZZ***
BZZZ-BZZZ-BZZZ-BZZZ—BZZZ-BZZZ-BZZZ-BZZZ

WHAP!

"FUUUUUUUUUCK!"

The Real Way to start the day: kill the fly buzzing its Monday reminder. Waking cry of existential dread. Bounce out of bed. The pretense of vigor quickly vanishes as the dizzying whirl from reclining to standing settles like dead weight reminding me I'm not fully awake no matter what my position. The rush toward the coffee maker remains a trudge, the attempt at speed forced but ineffective. Nausea: not so much at being alive as at my options. Do I:

(1) Say fuck it & go back to sleep?
(2) Call in well, stay home & write?
(3) Go to work, feel sick & write?
(4) All of the above?
(5) Some of the above in various permutations?
(6) None of the above?

The answer, as anybody knows, is indeterminate. The choices have no relation to the available options. In *Commercial Fiction* you experience the lives of the rich and the powerful, the beautiful and the enticing, the shakers and movers of business, government and lithe bodies.

So, what does going to work have to do with *Commercial Fiction*? You go where the money is. You experience the upper-middle class and the politically-indebted, the average and the obsequious, the sitters and squirmers of real life. Without them, you can't afford to write *Commercial Fiction*. They don't generate interest so much as material. Material reward for showing up, that is.

If showing up is 85% of life, I've already got B quality. B-movie without the tawdry glamour of *noir*.

Commercial Fiction is about showing up. And glitzing away the *noir*.

Before I trot out the Armani suits and the Lady Clairol, let me get myself ready for my 85%.

Let me get you ready too, by showing you the disparity between image and reality:

Image: The Author, hoary in his eminence, slowly sips his tea while thoughtfully reviewing the news of the day, as presented by *The New York Times*. Gently awakened, he seats himself at his desk and prepares to write.

Reality: The author, hairy with no spare time for a trim, runs between kitchen and studio, in various but increasing states of dress, while slurping coffee, printing out the morning's e-mail, making his lunch, cramming the printouts into his briefcase, grabbing his laptop and bolting out the door in a stress frenzy while the caffeine megadose kicks in prepares him for the road of insane drivers refuse to give me the right of way no matter how

important my writing about *Commercial Fiction.* Readers want *real* people, plausible characters they can identify with, living exciting lives in suburban mansions, not in urban ghost towns begging for new residents. The rich and the powerful don't live the latest incarnation of ers, junkies, alcoholics and

Insurance City Security Systems

1-800-911-1911

these are not the people hopes to attract to Insurance City's western burbs, where the latest generation of Stepford wives buys

Insurance City Condominiums

Downtown Living at its finest

starting at $99,999

behind the billboards above apartments housing hook- women who got it on with

COUNTRY MUSIC 96.5 FM

loud on their stereos, not to mention all the others. No,

Swarthmore & Ivy Clothiers

power suits and sweats.

Good Morning, Audience

(an interactive news program that demands author and reader participation. Its HOST *is a low-key nondescript white male who exudes a comforting combination of homey warmth and intelligence. The* HOST*'s image can be any morning news program host in television history, from Dave Garroway to Bryant Gumbel. His thick voice carries the mildly sluggish tone of a person waking.)*

Good morning, audience, and welcome to today's show. I'm pleased to see that our live studio audience is awake.

(Laughter. The camera pans the red-eyed crowd, then zooms in on a pudgy man in his mid-forties wearing coke-bottle glasses and a fire sale suit. The wiry protrusions from his overgrown crewcut and bushy mustache give him the appearance of an untrimmed schnauzer. The camera zooms to the button on his left lapel:)

[] [] []

(Aerial view: suburban tract homes. VOICEOVER of America's housewives:)

Housewife #1: Willya lookit that schlub!
Housewife #2: He may be a schlub, but at least he shows up.
Housewife #3: So does the furnace man. He keeps me hot, too.

[] [] []

(Studio. The HOST *continues.)*

On today's show we'll be bringing you the world news, the local news, the weather and our usual line-up of celebrity guests. In addition, *I'll* be bringing you something seldom seen on television. Is everybody ready?

(The audience howls its approval.)

We'll be right back after this station break.

[] [] []

(Medium shot, the schlub.)

Hi. My name is Ralph Putz, and I'm a Role Model. Now, most of you are used to seeing celebrity Role Models, people like Michael Jordan, Muhammad Ali, Mark McGwire and Harrison Ford, just to name a few. They do public service announcements like these or come to your school to tell you to stay in there and don't do drugs. Now, the truth is, most of you will never be celebrities like them. Very few of us have the God-given talents that will make us celebrity Role Models. I didn't, but I learned that I could be a Role Model just by using my own God-given talents. When I was your age and in your grade at school, I learned that my greatest God-given talent was Showing Up. That's what I did in school, that's what I did in life. Showing Up meant I went to school. Showing Up meant I didn't do drugs. Showing Up meant I went to job interviews. Showing Up meant I got hired. Now, there are some wise guys out there who will tell you Showing Up is only 85% of life. Don't let them steer you down the path to trouble. Take it from me:

SHOWING UP IS 100%OF LIFE

Be there.

[] [] []

(Studio. HOST, *irritably:)*

I'm here, I'm here. And with today's news, no less.

Today an earthquake scoring ten points on the Richter scale struck the land mass formerly known as the Soviet Union, dropping Russia and its warring neighbors into the Arctic Ocean. An estimated three hundred million lives were lost. Also lost were the political ambitions of many Republicans, who had hoped to win public office this election by reviving the Cold War.

An airliner, as yet unidentified, struck the Empire State Building late last night, killing all two hundred of its passengers and causing a power outage from Montreal, Canada to Washington, D.C. The streets of all communities affected by the outage are overrun with looters toting automatic weapons. The Pentagon has informed the President that the crash may be the start of a massive terrorist attack. The National Guard was sent to the area, but apparently got lost in the dark. To aid our defense team, the nation's utility companies have agreed to octuple their rates, effective immediately.

In Europe, the nations formerly known as Yugoslavia have forgotten who they were at war with and have begun attacking anybody they see, including their own citizens.

Mukhdhup Muhammad, newly self-appointed spiritual leader of the Nation of Islam, has withdrawn his support from the Fruit of Islam, drawing protests from Gay Rights groups worldwide.

In the local news, a recent drop in this show's ratings has caused the network to replace me after today's show. In a related sidebar, my wife, who is suing me for divorce and custody of our three children, is having an affair with this show's producer. He's moving into my Malibu home.

You want to know about the weather? It sucks.

And now, I promised you something you seldom see on television. I know it'll be the last thing *I* see on televison. Is everybody ready?

(The cheerful faces of applauding members of the audience. Cut to HOST, *who holds a snubnose revolver to his temple. A startled hush from the audience. Then, their shocked murmur competes with the station's canned laughter.)*

Yeah, yeah. You can't believe it, all you folks out there in TV-land? Well, get with the program. This is real.

*(*PRODUCER *enters from the left,* STATION MANAGER *from the right.* RALPH PUTZ *walks down the aisle toward the stage.)*

Get back, all of you. I'm not afraid to take a few of you with me.

(Medium. Everybody stops in their tracks.)

Now for my last words. Barney, I could kill you right now, but I'd rather you find out living with Angie is a fate worse than death. And to Angie, my beloved wife: I sold our house for a buck to a street person this morning and I didn't pay up my life insurance. I wish you the best of luck, both of you. You may deserve each other, but you don't deserve my assets. Not a damn one.

Fuck you, network. Fuck you, sponsors. Fuck you, audience. Goodbye.

(The audience boos and hisses. HOST *sticks out his tongue and thumbs his nose. He pulls the trigger. At the audience's collective shriek, the screen goes black.)*

[] [] []

"DEAD AIR! WE'VE GOT DEAD AIR!" the Producer shouts, rushing frantically between the bloody desk and the blackened monitor.

"We've got a dead host, too," says the Stage Manager. He picks up a phone. "Damned prop! Somebody dial 911."

"Never mind that. Somebody get a fill-in. Who the hell are *you*, mister?"

"Ralph Putz." Offers a handshake.

"Cut to a commercial."

[] [] []

(The Rock of Gibraltar, the logo of Gibraltar Life & Casualty, stands firm in center screen. Then, superimposed, an official document. A rubber stamp imprints "CANCELLED." The Rock dissolves. Cut to BANK REPRESENTATIVE *approaching the front door of a home radiating affluence toward the mover's truck parked in front. Behind the door, the sound of crying voices.*

Before entering, the REPRESENTATIVE *grins at the camera. The* HOST *is playing the part of the* REPRESENTATIVE*. Cut to interior:)*

Representative: I'm truly sorry that this happened. But you have to understand our position in this matter.
Host's Widow: But we have nowhere to *go.*
Representative: I'm sorry.

(Sobbing sounds as REPRESENTATIVE *leaves, grinning. Lamps and bureau drawers fly through windows in his wake. A bed and box spring roll down the sloping front lawn.* VOICEOVER:)

Gibraltar won't erode if you don't.

(A rubber stamp imprints an official document: "PAID BACK." *Cut to the Rock of Gibraltar.)*

[] [] []

(A seamy urban side street. Street people with stringy, unwashed hair lean against a white, shaker-shingled building. A gray-haired man wearing a week's supply of clothing under a wrinkled raincoat passes a bottle toward the other men. After sipping, they turn to the camera.)

Man #1: Hey, baby. Got a buck on ya?

(The camera moves forward.)

Man #2: Hey, lady. I'm a buck and I'll get on ya for twenty.

(The camera moves forward.)

Man #2: *(voiceover)* That's the best deal in town, hey.

(The WIDOW *and her children tread nervously between the men clustered around the entrance. Interior. The* HOST, *with a priestly demeanor, speaks to the camera.)*

I understand. The loss of a home can be truly devastating. Here at the Sister of Mercy Shelter we understand your pain. But we also understand the pain of people who had no home to begin with, of mothers whose children were born and raised in dumpsters. I'm afraid our beds are full.

(Weeping off-camera. Then, a slurred, raspy voiceover as the WIDOW *and her children trudge into the distance:)*

At Sister of Mercy, we help people who want to help themselves. If you want to help yourself, get here early. Or go sleep somewhere else.

[] [] []

"Let me get this straight," says the Producer, pawing back his thinning hair. "You're telling me that you want to just go up there in front of the cameras and take over. Right here, right now?"

"I'm here." Putz's shaggy mustache shadows his grin.

"We need someone with experience. Tony. Any luck?"

"Holly's in Hawaii, Phil's in Pennsylvania. Everybody else is an answering machine."

The Producer's appraising stare takes in the Role Model's paunch rolling over a low-slung vinyl belt. "You're right. You're here."

A Life in the Day of (2)

I'm here now, boxed inside the illusion that the physical world is an illusion and vice-versa. But I'm not sure which box I'm in: home or work. If it's my desktop computer I'm at home. My laptop I'm at work. Either place I think Outside the Box, the latest corporate buzzword for creative thinking in uncreative environments. But where am I?

Here's the walking Reality Check: Bubbles LaFlamme, the miniskirted mid-30's bottle blonde, parading her killer legs down the row of cubicles. Her four-inch heels click, even on the institutional carpeting. Male heads turn so fast coffee splatters their laps. Strange suppressed mating dances ensue. As she curves past my desk I notice her hair is shading toward platinum. "Your hair looks really bright today. You catch a lot of sun this weekend?"

"Only my hairdresser knows for sure." The Lady Clairol line through her kewpie-doll grin, a giveaway tic at its corners. Great looks, but no originality. She struts away from me, her gum snapping disco syncopation.

There's one box I'm not in. So I think creatively—and Outside the Box.

[] [] []

(Black screen. The sound of high heels clicking and gum snapping disco syncopation. At the center of the screen, a tiny dot appears. It hovers to the beat, emanating a golden glow. The hover becomes a spin that launches the

dot forward. The spinning dot enlarges and takes on the shape of a box. The box springs forward, a runway cage with a curvaceous woman dancing as wildly as a go-go girl, engulfed in the nearly blinding glow. A sharp pop of gum stops the music and the box. Under the FOX IN THE BOX *label* BUBBLES LAFLAMME *poses enticingly, hands on the hips of her leopard-skin bikini. Her glittering aqua gaze mists toward the camera. She speaks:)*

Are you Out of the Box? I'm not. But if you are, there's still hope for you.

A little dash of FOX IN THE BOX gives you that something extra that catches the eye of the shakers and movers in today's world. If you want free penthouse suites, luxury cars and handsome men who take you to *all* the right places, you want FOX IN THE BOX.

(The rhythm resumes. BUBBLES *dances wildly while the box spins back to a dot. A sunlike glow remains onscreen.* BUBBLES *in voiceover:)*

FOX IN THE BOX gets you where you want to go.

[] [] []

Straight out of morning TV commercials, that woman: self-improvement through beauty aids. Always the aura of mystery: the Mystery of Product, not the Mystery of Woman. If you've got it, flaunt it, I say. And Bubbles does.

"Ms. LaFlamme, Commissioner Johnson-Barr would like to see you in his office, ASAP," Melissa announces over the morning's first soap opera broadcast blaring from the mini-TV on her desk. The soft-eyed Clerk II, who considers dreaming of a modeling career her real job, resumes her sporadic keyboard clicks behind me, a tap here and there between the cast's lines.

Bubbles' hips sway over her heels as she approaches the elevator, a little extra strut in her stride.

[] [] []

(The commercial's rhythm. On the box, BUBBLES *dances into the distance. Stops. Looking over her shoulder, she says:)*

And *you* don't get in *this* fox's box.

[] [] []

Did that really happen?

Nice little spin, the commercial's personal touch. And while she's parading off to see the Commissioner's johnson bar, Bubbles is still sitting at her desk talking on the phone about yesterday's soaps: "I really wouldn't say he's an *admirable* character."

If I base her personality on a trace of conversation, I can give her bonus points for intellect; a college professor who taught the Modern Novel used to ask his students if such-and-such a character was admirable. It didn't seem like a very postmodern question to ask. But maybe Bubbles isn't as uninhibited as her image suggests. When a literary magazine called *Nude Beach* published a story of mine, I forwarded her name to the editors as part of my mailing list. The flyer they sent her came back unopened, "Take Me Off Your List" written in clear bold script. Since she didn't open the envelope I assume she was rejecting nude beaches, not me. She rejected me with the commercial come to life.

But did it really?

It's so hard to tell. The accumulated inundation of stimuli seems to create a spontaneous Virtual Reality. Bubbles is a fading advertisement, an elevator riding to the top floor, a telephone voice, a defiance of the conventional logic applicable to time and space. If she's in more than one place, then I'm not sure I'm at work. Maybe I'm home, typing in my morning fog, waiting for the caffeine jolt. But maybe I'm not.

Kyle the walkman'd stockroom clerk pushes an empty dolly past me. His unwashed ponytail flaps against the back of his black Joe Camel T-shirt. *YOU ARE WHAT YOU ADVERTISE.*

[Reminder: Order a T-shirt with your face printed on it.]

It's a violation of my principles, pushing somebody else's product with my chest and back. Free advertising. Slave wages. Right after the sixties I tried to explain this to the friends I made when I first moved to Insurance City. They *drank* Bud, so what was *my* problem? The same then as now: that youth rebellion comes down to wearing a commercial and listening to bad rock.

But Kyle's walkman overflow tells me I'm at work. He isn't someone I have fantasies about. Aside from delivering copy paper, he exists to sort my illusion from my reality. So here I am in the office, locked to the laptop locked to my desk. So many distractions here. Let's return to

—WE DISRUPT THE "FLOW" OF THIS FRAGMENTED NARRATIVE IN ORDER THAT WE MAY BRING YOU THE FOLLOWING INTERRUPTION—

Author: *(offscreen)* Come on. I'm in the head.

Voice: *(offscreen)* In TV we have a five-second tape delay. In literature we have an indefinite tape delay.

Author: When you do this, you make me feel as though I'm not in control of my material.

Voice: It depends on what you mean by "you." In a very real sense, I *am* you.

Author: True...

Voice: Not to mention the issues you raise by "control" and "material."

Author: This isn't exactly the time to wax theoretical.

(A toilet flushes. The AUTHOR *appears, pulling up his zipper and notching his trouser belt. He glances at the screen. Momentary shock, then blush. Through a wan grin struggling toward poise, he says:)*

Being a writer isn't just sitting down and telling a story or telling a non-story or not telling a story or not-telling a non-story and all the attendant ramifications of whatever theory is in fashion at the moment. Being a writer carries a certain responsibility. If you're attuned to what you're writing, the things you write can come true. It started happening to me in college. I'd type some situation, something *surreal*, something *totally* improbable, and it would come true. At least part of it, anyway. And it *still* happens. William Burroughs has written about it. When I read that he found things came true after *he* wrote about them, I didn't feel so isolated, even though the greatest part of my life takes place in isolation. But what I learned from having my own experience confirmed is that the writer, even in this high-tech era, occupies a shamanistic role, sometimes as shaman, sometimes as prophet, sometimes

—WE DISRUPT THE "FLOW" OF THIS FRAGMENTED NARRATIVE IN ORDER THAT WE MAY BRING YOU THE FOLLOWING INTERRUPTION—

Only this time I wasn't, uh, indisposed. I'll pick up the topic again later, if I can. If not, well, how much *real* closure does life offer before the Reaper beats us? Let's return to

Good Morning, Audience

The Producer flails a sheet of paper like a flyswatter against a swarm. "*This*—this is the most outrageous contract I've ever seen. If you think I'm signing it, you're crazy." The blue-purple blue-purple throb from the vein on his boiled temple coronas his balding head.

"Sane or crazy, I'm here, and you've got Dead Air." Putz is unflappable.

"What are you trying to *do* to me?"

"To you, nothing. I'm trying to Be All That I Can Be."

"Then join the *Army*, for God's sake."

"Been there, done that."

"Go there and do it again, for all I care."

"I already did. Would you rather I got an agent?"

"*You!*" Snickers. "You wouldn't know *how* to get an agent."

Putz grins his Schnauzer grin. "I know how to show up at an agent's office."

"And what would you do when you got there?"

"I dunno. The rest just seems to take care of itself."

A sinking expression consumes the Producer's face. He sags into the nearest seat. "I'll sign."

[] [] []

(The homes of America, aerial view. PUTZ'*s low-key murmur serves as*

background music.)

Housewife #1: He's not exactly handsome...

Housewife #2: But he's not really ugly...

Housewife #3: I like the way he stands up for himself.

Putz: I'm the morning host, not an undertaker. Please remove this body from my chair.

Housewife #3: See what I mean? He don't take nothin' offa nobody.

Putz: That's very rude.

Housewife #2: I'm beginning to see.

Putz: If you don't, I'm walking off the show.

Housewife #1: Just like Waldo when he puts his foot down. I *love* it.

Putz: That's better.

Housewives #1, 2 and 3: He's the *best*.

[] [] []

PUTZ TAKES OVER SHOW IN CRISIS

Emergency Walk-on Becomes Audience Favorite

It was do or die.

The air was as dead as the long-time anchorman filled the dead anchor's shoes after hurried negotiations with station low-key stage manner defused the impact of the tragedy broadcast only

[] [] []

(Interior. Plush office of the Network CEO. Glass cases holding rows of Emmies line the wall behind the PRODUCER, *who sits in a thickly-cushioned black leather chair that seems to dwarf him. The* CEO, *out of view, speaks with a booming authority whose every word pummels the* PRODUCER:*)*

YOU MEAN YOU CAN SIT THERE LIKE THE GODDAMNED FOOL YOU ARE AND TELL ME YOU ACTUALLY *SIGNED* THAT PIECE OF PAPER?

[] [] []

(A seamy urban side street. The PRODUCER, WIDOW *and* CHILDREN *weave around the homeless people crowding the sidewalk. Cut to interior: a cramped office with the backs of chairs forming a half-circle in front of the director's desk that faces the screen. Behind the desk and to its right, another desk faces the wall. In-baskets on both desks overflow with paper. Flanked by a pair of high-piled baskets, the* SHELTER DIRECTOR, *a woman with short graying hair and a severe face, introduces herself, then:)*

And this is the monitor from our funding source, Mr. au Thor.

I glance toward her clients, grin and turn back to my laptop. No point to correcting her Franco-Nordic corruption of what she thinks is my name. Nor to explaining that my name is really my function. Above all, no point to blowing the cover for my function: typing up my analysis of the shelter's operation.

"Mr. au Thor," she continues, "I hate to interrupt your work, but I am, as you know, a firm believer in the shelter service consumer's right to confidentiality."

"No problem. I'll come back and finish up after the interview."

I close the laptop on the back desk, leave her office and drop myself in the seat next to Vanessa Caliente, her secretary. Vanessa's beehive of thick black hair tilts back so that the overhead light puddles gently in the hollows below her cheekbones. She chirps click-click rapid-fire Español into the telephone. Her left hand twists and fondles a plastic lanyard woven from the colors of the Los Locos gang the same teasing way it will my cock. Her dark eyes dart in my direction, glitter black gold into my gaze. Once our eyes lock, she turns away. My eyes follow the long thigh curving through her skintight denims. She's teasing me every way she can before she shares with me the Shelter Director's latest injustice and her desire to replace her boss. Finally she clicks the phone onto the receiver and spins her beacon grin toward me.

"Hi, Arthur!"

(Game Show. Three CONTESTANTS *stand behind podiums.* CONTESTANT #1 *is a bony white male with long sandy hair matted down to his shoulders, four front teeth missing and a week's unwashed clothing layered over him.* CONTESTANT #2 *is the Producer.* CONTESTANT #3 *is the Widow. The* SHELTER DIRECTOR, *wearing a gray business suit, is the emcee.)*

Director: Welcome to *GIMME SHELTER*, the only morning game show that gets you through the night. And now I'll introduce our three contestants. Contestant #1, tell our TV audience what you do.

Contestant #1: Baby I do *lotsa* things, separately and in combination. You name it, I done it. Booze, pot, acid, smack, cocaine, crack...

Director: I know one crack you won't do unless you win a free Lysol bath, sweetie. Contestant #2, what do you do?

Contestant #2: I'm a Producer. Or was.

Director: If you win, sweetie, you'd *better* produce. Contestant #3, tell us what you do.

Contestant #3: I'm...a mother.

Director: Oh, you're a *re*-producer. Or former. You look like your trophy wife days were a long time ago. Of course, if you win, you *could* be a trophy *matron* for the night. *(Curls a finger under #3's chin.)*

I follow Vanessa up the stairs, the better to observe her cheeky sway. She unlocks the door to the family bedroom, lets me enter first, then presses the door shut with her shapely seat cushion. "You should *see* what she does, that Director. What I told you before? She talks fresh to everybody. She makes the people have sex with her or else they go sleep in the dumpster. She hits on the womens too. Even me. I don't like no womens hittin' on me."

I'm concerned. If she doesn't stop talking we might experience coitus interruptus when the family takes the room. "You like sex too," curving my arms around the taper of her waist and pressing my groin against hers. Instant hard-on.

"But we want different things."

Instant soft-on. "We do?"

"Not you and me. Me and her. She wants sex, I want to be Director."

"If you get it, don't talk to your clients the way you talk to me."

"What you mean? I talk nice to you."

"You have ulterior motives."

"Don't talk to me like that. I am not ulterior to nobody."

"That's not what I meant."

"I know. Sometimes we say things we don't mean."

"Maybe we should just stop talking," I suggest as she presses against me again.

"Mmmm..." through lips tasting mine one lip at a time.

We grind toward the double bed.

(Game show.)

Director: And here are today's prizes. Winner #3 gets to sleep on *(voice-over) AN ARMY COT!*

(Cut to army cot. VANESSA, *wearing a neon blue string bikini and glitter-gold high-heels, bounces up and down in positions exotic dancers use.)*

As you can see from my charming and beautiful secretary, this cot has the lively spring of a trampoline. If you bounce in your sleep, this will satisfy your every need.

And Winner #2 will win a wonderful and solitary night's sleep on *(voiceover) A SINGLE BED!*

(Cut to bed. VANESSA *poses on her side, the curve of her hips accentuated by the top leg stretching toward the screen.)*

Looks good, doesn't it?

---

feels good ohhhh baby the way your tight little pussy squeezes my rod omigod I don't know if I can hold come on hold back hold those lovely curves the soft belly flesh pressing against me the swivel of your hips those legs kicking the air rubbing down my thighs and calves your long nails digging into my ass you look omigod your hair spread across the mattress baby-faced Medusa nipples stiff brush my chest my cock oh god my cock the way I slide inside your softness satin silk no smoother than both oh oh ohhhhhh baby you whimper I moan the bed rocks

---

(Game Show.)

Director: And now for the Grand Prize! Today's Winner #1 will get to share—
All: OMIGOD!

*(*AUTHOR *and* VANESSA *spring off the double bed, cross their bodies with their arms. They scamper around the floor, snatching up clothing and trying*

to pull it on with little success.)

Director: Let's go to a commercial.

(Replay of AUTHOR *and* VANESSA *springing off the bed. Cut to a hype-meister in his late forties, raccoon-eyed, weak-chinned, wearing a discount store suit and a black toupee showing more loose threads than hair.)*

LIKED THAT MATTRESS ACTION, DID YA? WELL, HERE AT INSANE ERNIE'S MATTRESS PAD YOU DON'T HAVE TO CLIMB OFF. INSANE ERNIE HAS MATTRESSES THAT SUIT YOUR EVERY NEED AT PRICES THAT WILL MAKE YOU WANT TO ROCK AND ROLL. WE HAVE SINGLES FOR MIDGETS AND SOLITARIES, DOUBLES FOR COUPLES, QUEENS FOR THE DRAG SET AND KINGS FOR THE GROUP-GROPERS. WHENEVER YOU FEEL THE NEED TO DO THE DEED, JUST COME ON DOWN TO THE MATTRESS PAD WHERE THE PRICES ARE *INSANE.*

(Rates start at $29.95 for first four hours. Ceiling mirrors not included.)

Voiceover: TRAGEDY STRIKES THE AIRWAVES.

(In slow-motion, the host of Good Morning, Audience *shoots himself and falls across his desk.)*

ONE STAR FALLS, ANOTHER RISES.

(PUTZ *sweeps a gesture of revulsion at the body on the desk.)*

ALL THIS AND MORE ON ***THE NEWS AT NOON.*** BE THERE.

(Voiceover:)

I'm here, I'm here. *(*AUTHOR *turns away from laptop to grin at the screen.)* In today's fast-paced high-stressed world it's more and more difficult to separate factoid from fiction. You didn't think I'd *really* be dumb enough to get caught with my pants down, do you?

(Medium shot. AUTHOR *looks at shirttails dangling over his bare crotch, a white sock on his left foot, a Reebok on his unsocked right.)*

I guess I'd better get back to work.

While the shelter's night-time use is down from the projected levels, a recent field visit indicated the shelter has enormous potential for daytime use. It is my recom-	**Ctl** + **F6**	And the winner is... Contestants#2 and 3. You get to join me in a *menage a trois*. Oh, you won't? Sorry, no consolation prizes. The winner is Contestant #1. So,	**Ctl** + **F6**	When I'm locked to my laptop time and space assume a life of their own, both inside my head and outside in the real world. Sensory bombardment blurs the line separating

(The seamy urban side street. The PRODUCER, WIDOW *and* CHILDREN *trudge through the homeless people crowding the sidewalk. Each taunt starts one of them sobbing. When they pass out of view, the* HOST *steps into center screen, smirking:)*

In every game, there are winners and there are losers. *GIMME SHELTER* makes room for both.

(Turns and walks away *from the camera. Offscreen a coarse voice cries out:)*

This is *my* goddamn dumpster. You take your bitch and them brats an' go find your own. Bald-headed bastid!

[] [] []

Yeah, the hair's still there. But where was I? Somewhere...struggling to sort my senses from the stimulation they receive. Even a 4-F like me wants to Be All You Can Be. But are Reaching Full Potential and Information Overload the same thing? Even when I'm locked into my laptop mind body and soul the communion of myself with the words that portray my perceptions on the screen interior reality expresses itself as nonstop bombardment by exterior reality or what used to be exterior reality. Even though I live in my head like most writers the stimuli rocket through endlessly violating my sensibilities imagine a woman raped repeatedly no pause for torn vaginal tissues to heal. A spear of deodorant stick here a dick of toothpaste there all guaranteed to aggravate my bleeding hemorrhoids despite their money-back guarantees to keep me odor and plaque free through the next seventy-two hours of media attacks. Even Your Best Friends Won't Tell You Your Breath Stinks Your Armpits Reek Your Crotch Rots. But the heat-seeking missile of Truth strikes the minute you stare into the bathroom mirror: the tube of toothpaste blasting itself through the other side of the mirror its mint-flecked paste spraying the glass my reflection and me like the air escaping a balloon, the deodorant stick springing to scented life rubbing my armpits so raw that its alcohol base seeps into the broken flesh and burns it. Even fast-forwarding through the commercials is no escape. In the Golden Age of TV one commercial separated Sid Caesar from his next skit or the next round of the Gillette Cavalcade of Sports Friday

Night Fights. Then there were two then four now eight. When I hit the remote's play button I think I'm back to watching the show but the ad runs like a narrative until the protagonist antagonist or enabler whips out the product and the action-packed adventure climaxes with a guarantee of safety against graying hair dirty laundry sink stains body fat or impotence. The mechanism for co-opting dissent that began in the fifties and turned the protests of the sixties into commodity quests has reached new levels of sophistication

BIG BROTHER IS NOT WATCHING YOU
BIG BROTHER LIVES INSIDE OF YOU

but there are always the nameless Winston Smiths like me who try if not to subvert the message then at least to let others outside the reach of my immediate consciousness know that they're stuck playing a shell game. Turning the products back on themselves inverting their cliches even becoming them with a vengeance, these flailing strategies at least offer resistance to

Big Brother: *(on intercom)* Get out of your head and get your ass in here. I've got an assignment for you.

[] [] []

"Your Breath Stinks Your Armpits Reek Your Crotch Rots," Big Brother says through his trim black mustache, his long finger stabbing the air in front of me.

"Your knuckles need Hair Remover," I retort.

"And this shelter report needs Blank-It, the Thorough Whitewash. Not a trace of stain in twenty-four hours or your money back."

"I didn't even realize I'd completed it. Last I knew it was my cover document while I was working on *Commercial Fiction*, an attempt to express reality by using the imagination to extend its insidious reach. Do your thing till someone comes then switch the screen. Ctl + F6: *FAST!*"

"Man, you move so fast you can't even keep up with your*self*."

"What do you want, Bro? A ritual diatribe against the fast-paced Modern Age that makes each of us a walking Stress Attack? Every generation says that about its technology."

Big Brother leans back and laughs. "That's not what I want. I want to know about some alleged irregularities at the Sister of Mercy Shelter."

"What do you want to know?"

"That young babe, Vanessa. How is she for sack action?"

"I've never tried her."

"That's not what she says. She told the Commissioner's office she's filing sexual harassment charges against you and against the Shelter Director. She says there's a big game going on."

My fingers wipe the sweat off my forehead. "I thought I was just imagining that." Wasn't I?

"You gotta watch that," Big Brother says, lighting a cigar in his Smoke-Free office and switching on an air-filter the size of his new computer monitor. "People imagined electricity, automobiles, airplanes, spaceships ...and you know what? They've all come true. As a public servant, you've got to be particularly careful. Remember when Jimmy Carter was running for President? He told *Playboy* he lusted in his heart. That's the same thing as lusting in real life."

"Come on! That's just the old puritanical guilt trip."

"It's also the logical extension of ending the mind-body duality."

"I guess I should be more aware of unintended consequences. So, what's the *real* problem?"

Big Brother leans back in his chair. "There's a guy on the Shelter's Board named Putz..."

"Don't tell me it's Ralph Putz."

Brother's backhand sweep shakes the ashes off his cigar's inch-wide cinder. "Yeah, that's the dude's name."

Damn! This is getting spooky.

"Why? You know him?"

"A little. By reputation, mostly."

"You'll get to know him a little better today. The Commissioner wants one of my staff to meet with him."

"And I'm nominated."

Big Brother grins. "You're Number One."

Mr. Putz Has Left the Building

the receptionist tells me the celebrity cliche same tone every time except after the fifth she drops the "I'm sorry" preface her polite smile fades to a half-grin then tightens to a sneer almost a snarl with each repetition.

"Look." I rest my hands on her desk, lean across its institutional green blotter. "I'm not trying to be a pain in the butt—"

"If you're not, you must have a great natural talent for it."

"He made a call to Commissioner Johnson-Barr this morning. I'm supposed to meet with him about a matter of great concern to him."

An expression of resignation shades her face. At last! I've worn her down. My writer's persistence in the face of insistent failure has paid off in *one* area of my life, at least—though not the one I want it to. "Why don't you go through the door on the right and have a seat? Someone will be with you shortly."

"Thank you." Victory's in my grasp, I can tell.

But you couldn't tell that by the cubicle I'm suddenly sitting in: walls so close they remind me of the cells in horror flicks whose walls crush their prisoners and a light so blinding I wonder if I've stumbled into a brainwashing compound. My elbows bump the walls every time I move. What feels like a long period of silence under the lights in the cramped booth disorients my senses. Then a speaker crackles and a voice asks: "And what is *your* reason for being here today, Mr. X?"

"I'm looking for someone."

"You're here looking for That Special Someone?"

"Special to some people, I guess." What more can I say about Putz? I never heard of him before he showed up this morning.

"And if you were to find That Special Someone today, what would you say?"

"Let's try to work things out."

"You must be taking lessons from Alan Alda. Tell us, Mr. X. How would you describe yourself, sight unseen, to That Special Someone?"

How *would* I describe myself? That's the hardest thing to do. Middle-aged, graying, physically fit despite my beltline paunch...Too factual, too mundane. A writer trying to dig past an evanescent reality, hampered yet helped by the internal and external derangement of his senses? Too abstract. A pencil-pusher who runs his writing reconnaissance covertly on company time? No, gives away too much. Wrong context, anyway. A government flak-catcher? No. Putz would know that already...

"I see. You're a strong, silent type," the over-amplified voice booms and echoes around me.

My hands raise to protect my ears. My elbows bang the tight walls.

"Aha! A shaker and a mover. We have a strong, silent type who's also a shaker and a mover. It's time to make your choice."

"What choice?" I ask. The silence tells me the man hasn't heard my question. The heat and light are making my forehead sweat. I reach into my left pocket, pull out a handkerchief. My elbow whacks against the wall. I wipe my forehead. Thump the wall again.

"You'd better make up your mind. The strong, silent shaker and mover is rattling his cage."

My ears! Dammit! Both arms rattle the cage.

"Congratulations, Mr. X. You have been chosen as a winner on this portion of *The Mating Game.* Come out and meet your new partner."

The door opens. I step out of the booth, my eyes still filled with blazing light, stumble and fall flat on my face. Laughter. And another voice booming around me:

"TODAY'S WINNING COUPLE WILL RECEIVE AN ALL-EXPENSES PAID TEN-DAY VACATION AT THE SOUTH POLE MOTOR LODGE IN ANTARCTICA, A 1967 FORD FAIRLANE AND A TEN-DOLLAR GIFT CERTIFICATE FROM ACME TOWING COMPANY OF COLUMBUS, OHIO."

Winning couple! *The Mating Game!* What did I stumble into? My vision clears enough for me to make out Bubbles LaFlamme's waxed legs and high wave of platinum hair gleaming under the studio lights. Ten days with Bubbles at the South Pole! Those long legs lovelocked around me, those full lips brushing mine! And the penguins her only alternative!

Yes!

"NO!" Her eyes blaze green flames. *"NO! NOT HIM!* ANYONE *BUT HIM!"*

The camera tracks her hysterical exit. She presses the elevator's down button, glares impatiently at the floor indicator. Takes the stairs instead. Her four-inch heels break. She tumbles down two flights of stairs. Rises, hair disheveled and darkened with stairway grit. Runs outside. Climbs into the rusted-out Fairlane. It doesn't start. She leaps out, slams the door. Flags down the Acme Towing truck. The brawny driver, tanktop ponytail and tattoo, offers her a hand. The camera zooms in on her derriere swinging up the steps and onto the seat next to him. His arm wraps around her shoulder. The truck drives toward the horizon.

Next thing I see on the monitor is me: a gape-mouthed blush sagging at shoulders and knees.

[] [] []

HOW WILL OUR HERO ESCAPE THIS EMBARRASSING DILEMMA? WILL HE ENJOY A TEN-DAY SOLO STAY AMONG THE PENGUINS OF HIS CHOICE? WILL HE TAKE VANESSA, THE DAY-BED WONDER OF THE SHELTER WORLD? OR WILL OUR WOUNDED HERO

switch screens? Why not? Another reality awaits its creation and channel-surfing has its own analogues, *deus ex machina* be damned

leave the building Putz has already left before I stumble into a crazed rumble some shockjock host started on his talk show to boost ratings

report to Big Brother that I've failed in my mission not that he gives a damn beyond keeping a fresh stash of big Cuban cigars

[] [] []

assuming I was ever out of the office in the first place. In a network of simultaneous realities I can surf any channel. Maybe I've had my morning fitness workout with Vanessa Bedgrave, my rejection immunity injection from Bubbles LaFlamme and my daily dose of philosophical inquiry with Big Brother. So, where am I now? Which is illusion, which is reality? Any button I care to press will take me somewhere equally present, equally lacking in real closure. Continuity becomes the sequence of juxtapositions experienced, history the juxtapositions remembered.

How to separate illusion from reality?

Illusion: The author takes his morning break from his day's creative labor. Makes green tea, turns the TV to PBS. Misses the button on his remote. Instead finds himself watching Psycho Babel, the controversial talk show host whose raging issue today is Reverse Transsexual Gender Abuse Among Middle-Aged Necrophiliacs. The vehement reactions of the guests pique his curiosity. He watches so intently that he doesn't feel his fingers blistering from the heat of his Pfaltzgraff cup. Very interesting...in a vulgar sort of way, naturally. But...what is *this?* An image of himself, graying and wearing a corduroy jacket, stumbles into the screen, a perplexed look on his face.

"Who are you and what are you doing here?" Babel demands through a predator's snarl.

"It's an accident. I was looking for a Putz that, uh, left the building."

"You want a putz?" snarls one guest, a biker type whose broad shoulders bristle. "Talk to the surgeon that didn't stitch mine back on this last operation."

"Maybe it got bitten off at your last funeral home. Post-mortem reflex," says another.

What am *I* doing here? the author asks. "It's not *my* putz," he says. " It's a *person* named—"

"I don't care what you're looking for," the biker snarls. "I was talking and you interrupted me. That's very rude. You know what I do to rude little punks like you?"

Before the author knows it, the biker has lifted his duplicate overhead. Through the screen he goes, landing face-first on the livingroom floor.

"Let's stomp the wimp."

The guests climb through the broken screen, kick the fallen author in the face and sides. The watching author backs away from five attackers. One lunges for him. In self-defense he hits the attacker with his remote. The channel button changes to a talk show panel discussing Spousal Abuse Among Masturbating Bachelors and immediately cuts to a commercial:

IS GETTING BEATEN SENSELESS RUINING *YOUR* COMPLEXION? THEN *YOU* NEED PAM'S POLYURETHANE PANCAKE MAKEUP. PAM'S POLYURETHANE PANCAKE MAKEUP LETS YOU TAKE THOSE TOUGH BEATINGS WITHOUT WORRYING ABOUT UNSIGHTLY BUMPS AND BRUISES. JUST ONE APPLICATION A DAY WILL KEEP YOU LOOKING LIKE A *LOVELY* TARGET.

(Caution: the Surgeon General has determined that Pam's Polyurethane Pancake Makeup is of limited effectiveness in preventing internal injuries such as cerebral hemorrhages.)

Reality: "The receptionist kept telling me he left the building," I tell Big Brother over the air purifier sucking up the yellow smoke that billows from his Cuban's glowing tip. "I tried."

"Don't sweat it. You were there, he wasn't. Let the dude call back."

"After that sexual harassment charge, I'm a little leery of any reality I walk into. Assuming they're realities in the first place, that is."

Brother leans back, his black mustache and gapped teeth grinning reassurance. "It's not the realities that you walk into that you gotta worry about. It's the realities that walk into *you* that you gotta watch."

"Good point." Brother's hardly an Orwellian projection. He seems to know the score as it's tallied in the back rooms of the world. He should; he manages the subliminal information programming that has shaped the culture."Fuck it. Let's take a break."

"It's closer to lunch than mid-morning."

"Did I say 'Fuck it'?"

"You did. At least in this reality."

"Then let's hit the cafeteria."

Over steaming coffee and the front page of the tabloid *Amerika Today* we swap small talk, glancing through the circular glass wall at the Interstate so close a car could leap a guard rail and land on us and at the nearby security desk vulnerable to end runs just like the one happening now guests from Psycho Babel's show breaking through the guards' TV hurling one frail badge through the screen sending him skidding across our table till his nose breaks against the view of the Interstate. The guests overrun the lobby then the caf pounding the employees who haven't fled to cover or ducked under a table. Finally they leave.

Brother looks up from under the table. "What'd I tell ya?"

I raise my head. "You're right. Looks like this one walked into us. It looked pretty damn real, too." I pull the *Amerika Today* off my head. What protection did I think it could offer me against these crazed Reality Invaders? I should just crumple it up and toss it in the "THIS IS NOT A WASTE BASKET" waste basket that reminds me of the seminal postmodern work, "*Ceci n'est pas une pipe.*" But I pause long enough to read its headline:

PUTZ WALKS OFF SHOW, OUT OF BUILDING

Rising Star Cites Network President's Lowball Offer as Reason

What does TV's hottest star do when he gets hot under the collar? The headline says it all. Ralph Putz, who was able to raise *Good Morning, Audience*—but not its former host—from the ratings dead in a remarkably short time, expressed his strong dissatisfaction with the network's latest offer by leaving the show he saved single-handedly. At issue was appropriate compensation for his incredibly rapid rise from a Role Model in public service announcements to the most in-demand emcee in today's competitive morning TV market. "I know what I'm worth," he

The News at Noon

(Newsroom. The male co-anchor sits on the left side of a kidney-shaped desk, the female co-anchor on the right. Each is carefully-groomed, about age thirty-five. The male is African-American, the female a blonde Caucasian. In a crowd of attractive people, neither would stand out.)

Ryan Bumble: Welcome to *The News at Noon*.
Jessica Candy: We'll be back with today's news after these messages.

*

Bumble: Some say he's the best
Candy: But none say he's the brightest.
Bumble: Today Ralph Putz, the rapidly-rising star of *Good Morning, Audience*, walked off the show he had saved from certain cancellation earlier this morning.
Candy: At issue was the salary Putz received for saving the program after its former host committed suicide at the start of today's broadcast.
Bumble: Network President A. Barnum Rube refused to grant Putz's on-air request for $20,000,000 a year, citing the former host's suicide as the reason for the astonishing rise in today's ratings.
Candy: Putz, known for his low-key, affable style, asked his staff to return the dead former host to his former desk before walking out.

Bumble: Well, Jessica, I guess that's just his low-key, affable way of saying, "I'm mad as hell and won't take it anymore."
Candy: You know, that line from Paddy Chayevsky's *Network* calls to mind a related story.
Bumble: We'll show you after these words from our sponsors.

*

Candy: Early this morning, after receiving his pink slip, the former host of *Good Morning, Audience* took his life before an entire nation of viewers.

(Footage from the show.)

BBBB you, network. BBBB you, sponsors.BBBB you, audience. Goodbye.

(The audience boos and hisses. In slow-motion replay: HOST *sticks out his tongue and thumbs his nose. He pulls the trigger. At the audience's collective shriek, the screen goes black.)*

Bumble: That's one way to start your show with a bang.
Candy: *(A cold glare, then:)* Is there Life after Death? You'll find out—after these messages.

*

Bumble: You've seen him everywhere today.
Candy: You've seen him alive, you've seen him dead
Bumble: And you've seen him alive again.
Candy: By now, you're probably wondering "Is he or isn't he?"
Bumble: Talk about Elvis sightings! The former host of *Good Morning, Audience*, took his life earlier this morning at the start of his show.

(The suicide scene replays in extreme slow-motion.)

Candy: Despite his passing this morning, the host has been seen in commercials for Gibraltar Life Insurance and public service announcements for the Sister of Mercy Shelter. Our news crew spoke with the appropriate spokespeople earlier today.

(Front steps of Gibraltar Insurance. The CEO, *a white-haired man with patrician features, responds to the familiar-sounding voiceover.)*

Voiceover: These commercials were made prior to his death, weren't they?
CEO: It is my understanding that they were taped after he left *Good Morning, Audience*. We would certainly want to avoid any appearance of conflict of interest with regard to our hiring practices.
Voiceover: Are you saying these commercials were filmed *after* his death?
CEO: It would be inappropriate for me to comment on that.
Voiceover: Why would it be inappropriate? The man is either alive or dead.
CEO: Gibraltar sells Life Insurance. If people who rose from the dead were to collect on their policies, we would be out of business within a very short time frame.

(Entrance to Sister of Mercy Shelter. DIRECTOR *stands on the sidewalk in front of it.)*

Voiceover: Did you know he was dead?
Director: We serve so many people it's hard to say. Given the desperate straits they find themselves in, many of our service consumers look as if they might be dead.
Voiceover: When was the ad filmed?
Director: This morning.
Voiceover: Before or after he shot himself?

Director: In the mornings we hold our *GIMME SHELTER* evaluation, where we determine which needy applicants will get the beds that are in such short supply due to our inadequate funding.
Voiceover: Before or after?
Director: I really can't say.

(Studio.)

Bumble: But there are some who can.

(Street. The WIDOW *and the* former PRODUCER.*)*

Voiceover: Are both of you sure he was alive?
Widow: He was *definitely* alive. When we left the shelter, I looked back and...and there he was, big as life. Er, almost as big. And he had this *smug* little gloat.
Producer: He said he'd let me live so he could see me suffer with her. But I had no inkling whatsoever that he would find a way to actually stick around and *watch.*

(Studio.)

Candy: So, the answer to the question of whether there's Life after Death remains up in the air, where only the Higher Powers know.
Bumble: Meanwhile, down on the earth, a sex scandal is brewing at a local homeless shelter.
Candy: We'll be back after these messages.

(The HOST *appears. His voice is the* VOICEOVER *heard in the previous feature.)*

Is there Life after Death? Maybe, maybe not. I don't have all the answers. Am I alive or dead right now? I can't really tell you. But I can tell you what I use when I want to get out of a life I'm in. I use the WESTON-SMITHERS snub-nose .357 magnum revolver, complete with two-inch barrel and pearl handles.

(Revolver rotates slowly as if floating in space. Voiceover:)

The Weston-Smithers snubnose .357 magnum revolver guarantees maximum impact at close range. The optional hollow-point cartridge expands upon contact, ensuring terminal damage to all living tissue.

(Cut to HOST.*)*

I take mine with me whenever I want to go.

(His on-air suicide replays in extreme slow-motion. Instead of fading to black, the screen shows him lying across the desk, the visible side of his head disfigured, smeared with blood and brain tissue. Beneath him a small trickle of blood runs over the edge of the table and down its front. Back to HOST:*)*

Remember: with a Weston-Smithers, you can't miss.

THIS IS *NEWS!* GIMME A BREAK! NO, TAKE THAT BACK. I'VE SEEN ENOUGH BREAKS. BUT COMMERCIAL BREAKS FROM THE AFTERLIFE! LOOKS LIKE THIS GUY CORNERED THE MARKET. NEXT THING YOU KNOW, HE'LL BE DOING FUNERAL HOMES. I'D BETTER GET BACK TO

"Hello, Arthur? This is Vanessa. I know you're at lunch right now, but I need to talk to you about something very important. My boss, the director? She tried to make me have sex with her right after you left. I told her I don't do nothin' like that with no womens and then she told me I was fired and I had to leave or she would call the police. I need to talk to you. Maybe I could see you? Please call me. I'll be at home watching"

The Bald and the Ruthless

(The program's maudlin organ theme segues to the office of COMMISSIONER JOHNSON-BARR, *a fortyish male wearing a closely-trimmed full beard and a blue pinstripe suit.* BUBBLES LaFLAMME *enters tentatively.)*

Bubbles: You wanted me to come to your office?
Commissioner: Not exactly. I wanted you to come *in* my office.
Bubbles: Excuse me, sir. Are you prepositioning me?
Commissioner: To be frank...
Bubbles: I mean, I don't do prepositions well.
Commissioner: I'll bet you're great at conjugating.
Bubbles: At...What's *that?*

[] [] []

(Restaurant. Stock upscale set: an abundance of ferns, hanging plants and wood furniture. PUTZ *sits across from the* SHELTER DIRECTOR.*)*
Director: I think you can understand the problems she poses for us.
Putz: She definitely sounds like a loose cannon.
Director: It's not the kind of information you pass along to your funding source.
Putz: You had no choice but to let her go.
Director: Vanessa? Yes, she's not a good worker. Moreover...

Putz: You mean the rumors about her being involved in drug gangs?
Director: That, too. But what I was going to say is, she's not a Team Player.
Putz: That's even worse.
Director: It only takes one person like her to destroy a viable concept.
Putz: Even one with Network potential. The funding source is supposed to meet with me. I'm waiting to hear from them.
Director: I'd appreciate anything you can do for the shelter.
Putz: I'm not the kind of person who forgets where he came from.

[] [] []

(Commissioner's Office.)

Commissioner: For the special project I have in mind, I need someone who can work with me very closely.
Bubbles: I understand. It's a very big project, Commissioner.
Commissioner: You can call me Commish.
Bubbles: Thank you, sir—I mean, Commish.
Commissioner: In these situations I like to dispense with formalities. Now, about this project, I need someone who can work with me, not only closely, but intimately as well.
Bubbles: On a project of the dimensions you describe, you couldn't do it any other way.
Commissioner: That's the yardstick I go by.
Bubbles: It's a great tool for measurement.
Commissioner: I've never had any complaints.
Bubbles: And you won't hear any from me, Commish.
Commissioner: Before we wrestle further with this, I need to know just a little more about your qualifications. It's essential that my assistant have exceptional oral skills.
Bubbles: Well, gee, I mean, that is, I'm perfectly good in private. Orally, you

know. So...you know...I guess I could *learn* to be as good in public. I mean...
Commissioner: When the situation arises, I'm sure you'll address an important need. What about your other skills? Can you take dictation?
Bubbles: All I need is a pad.
Commissioner: *(Pulls out a black leather outfit with studs, a whip, handcuffs, etc.)* We don't need to worry about that. I'll tell the secretary to hold my calls.

*(*BUBBLES *looks at him with a vacant expression of submission.)*

[] [] []

(Seedy urban apartment. High ceilings, faded beige paint, cracked plaster. Soap opera organ droning in background. VANESSA *sits next to* PACO *on an old, sunken sofa. Each wears blue jeans and a black tanktop.)*

Vanessa: You got to *trust* me, Paco. That man, that woman...they don't mean *nothin'* to me.
Paco: Just make sure they don't, *chica.*
Vanessa: That Anglo Arthur, he back me to save his own butt. They fire her, they hire me back. As Director. Who else they got?
Paco: And Los Locos got a clubhouse.
Vanessa: You got it.

[] [] []

(Exterior. High-rise office building. Zoom in. Cut to plush interior: thick wine-red carpet, floor-to-ceiling windows looking out to the horizon. AUTHOR *stands at a circular desk, hovering over an indifferent* RECEPTIONIST *reading* Vamp *magazine.)*

Author: I'm here to see Mr. Putz.

Receptionist: Get in line.
Author: He's here, then?
Receptionist: He's here, but he's in the middle of some very important negotiations.
Author: My meeting with him is important too. Can you tell him I'm here representing Commissioner Johnson-Barr?
Receptionist: You'll have to wait. I'm going on break. *(Leaves.)*

(AUTHOR *goes to the double door behind the receptionist's desk. He listens:)*

Putz: I'm telling you, this is not *my* career move. This is *your* career move.
Female Voice: I don't see it that way.
Putz: I'll *make* you see it that way.
Female Voice: Don't threaten *me*.
Putz: Go ahead. Shoot.
Author: Oh, shit!

*(*AUTHOR breaks through the door. PUTZ *is naked. His beefy butt rises up and down as he fucks* PAMELA ANDALE, *a TV star known for her big blonde hair and even bigger silicon breast implants. A camera man films the couple while his crew handles the lighting. Above the massive waterbed a sign reads "TWO O'CLOCK MATINEE.")*

Author: I—I'm sorry. I mean, I didn't mean to—I—your secretary told me you were in a meeting.
Putz: *My* secretary!
Andale: Ooh! Ooh! Ooh! You must've talked to *my* secretary.
Author: Whoever's secretary she was, she told me you were in a meeting, not in a...uh...movie.
Andale: We *are* in a meeting. In fact, we just met.
Putz: I just showed up.

Andale: Pudzi's a *very* powerful man. One minute he's in the waiting room, next minute he's in here, talking turkey.

Putz: Talking turkey *neck*, she means.

Andale: *(small voice)* Ooh! You're such a naughty boy, the way you talk.

Putz: So, what did you barge in here for?

Author: I was looking for—

Putz: A job? There's a Role Model opening available.

Author: I was looking for you. Commissioner Johnson-Barr told—

Putz: Commissioner Johnson-Barr can wait. This comes first.

Andale: Not if you don't start moving it, Pudzi.

Author: Sorry to disturb you. Maybe some other time...

Putz: Yeah yeah yeah.

(AUTHOR *turns to leave. Behind him:)*

Andale: Remember, Pudzi. You promised.

Putz: I'm not the kind of person who forgets where he came.

A Life in the Day of (3)

This isn't the first time my life has turned into a soap opera, but it's the first time I've ever *walked into* one. It's the first time I've *ever* walked into the filming of a porn movie...or was that a soap opera in which a porn movie was being filmed? The disruption of time and space in contemporary media affects contemporary life and contemporary literature, as well. If continuity continues to exist as a linear concept it does so only because memory links one burst of sensations to another. I'm no longer sure whether I'm at home at work or in a soap. The $64,000 Dollar Question is: how do I find out?

Dr. Joyce Brothers stares at me from her cubicle, a sage glance from the 1950's when the show launched her into lifelong celebrity. She bends slightly forward from the waist to press a buzzer. "Actually, Author," she says, the first person to get my name right, "reality is composed of many elements, all of which originate with perception. It is possible for one person to have one perception of reality and another person a different perception, simply due to our individual differences." She smiles: a pleasant face, even teeth.

"Once again the winner!"the emcee shouts. The cameras turn away from me. Apparently I'm the loser on a show I never knew I was participating in. I step away from the hullabaloo surrounding the winner. Discover my office cubicle has shrunk around me. Now I have to exit through a door that wasn't there before.

On the other side Bubbles LaFlamme says to one of the female clerks in the unit: "You can't exactly call him an *admirable* character." Apparently her critical repertoire has its limits.

"But he's a Role Model," Melissa insists, continuing to type her five words an hour without distraction or damage to her three-inch nails.

"I think the Commissioner is a better Role Model than he is."

"Him!" Melissa stops completely. "They say he's only got one thing on his mind."

"But he *is* an attractive man," Bubbles insists.

"So, which of the two would you rather do a skin flick with?"I ask.

They glare at me. My mouth feels crowded with my Reebok in it.

- **Do *you* stick your foot in your mouth more often than you should?**
- **Do your heels scrape your tongue when you talk?**
- **Do friends you've never had stare at the shoelaces that dangle over your lips?**

Then *you* need

PUCKER ALUM

the mouthwash that really works.

"Oh, Art. It's so nice to talk to you. You're so different now."

"Uhm thwk wmm mm mfff."

"It's really *nice* of you to say that."

Remember:

PUCKER ALUM

"The mouthwash dentists recommend to their patients"

Was I watching the same soap opera that they were listening to on Melissa's desk radio? Or was I writing a soap of my own? More $64,000 questions and no Dr. Joyce to answer them. Not even her cubicle sticking out on the floor like a cut-rate gazebo. And what about the commercial? Which is illusion and which is reality? I can no longer tell. Even the world's remaining Winston Smiths are turning to Big Brother for answers.

Big Brother's office more and more resembles a mall cigar store. Behind its plexiglass walls the ferns concealing the NO SMOKING sign wither crinkle and die. Their leaves flake into the roiling yellow cloud pungent as cat piss.

"What can I do for you, boss?" Big Brother asks as I close the door behind me.

"I'd like a reality check."

"Wait till payday."

"How will I know that's real?"

"You don't. But that's as real as it gets."

"Well, I had this experience..." I tell him how disoriented I'm feeling.

"Wherever you are, you're here," he assures me.

"But where *is* that?"

"If you don't know, I can't tell you."

Even Big Brother has his limits in a media-saturated environment, I guess. I tell him about *The Bald and the Ruthless* in general terms. He leans back from his long puff and says, "You stepped in some deep doo when you stumbled into that porn flick, man."

"How did you know about that? You see the show?"

"No..."

"It must've been the radio blasting out there."

"No..."

"Then, how did you find out?"

"We live in an age of Instant Communication."

"I just mentioned two forms of instant communication."

"The Commissioner communicates even more instantly when you piss him off. He was in the Luv Tub Motel with Bubbles when Putz called him."

"But....Bubbles has been right *here...Hasn't* she?"

His gap-toothed grin flashes through a break in the smoke. "Sometimes you're too logical for your own good."

"What's that supposed to mean?"

"You keep thinking reality is consecutive, man. Maybe it's concurrent."

"Those sound like prison terms."

"They are. Concurrent gets you out sooner. But there's still no escape."

No. 2: You're Number Six.

No. 6: I am not a number. I am a free man.

No. 2: You're Number Six.

No. 6: Who's Number One?

No. 2: I'm Number Two.

No. 6: You certainly are.

Nothing like a flashback to *The Prisoner* show from the sixties to remind me of my youthful delusion that I could actually escape the mind-body prison.

WE'RE NUMBER ONE

WHO SAID THAT?

I can't even remember the commercial that appeared in. Maybe the failure of memory is the way to escape the system's all-embracing mindtrap. But the same failure of memory allows the mindtrap to exist in the first place.

"You wanna know who said that?"

I turn. Big Brother's standing behind my cubicle. How did he know what I heard? How did he get out of his office? For that matter, how did I? Does it matter?

"Here's the scoop." He hands me an Extra Special Edition of *Amerika Today*. Its headline reads:

TALK SHOW PUTZ BECOMES NETWORK PREZ

Is he a rocket, or merely a Roman candle? Only time will tell. Meanwhile, Ralph Putz continues his meteoric rise from a mediocre role model to a mediocre talk show host and beyond.

Putz, who walked off his hugely popular *Good Morning, Audience* show because of a salary dispute with network executives, used his instant popularity to leverage a position industry professionals would have considered inconceivable.

Until today, that is, when the Board of the International Broadcasting Company surrendered to Putz's hard-line strategy and gave the role model turned celebrity host, in effect, *carte blanche* to oversee not only his own show, but the entire network as well.

"Looks like our home boy is making good," Big Brother says.

"And redefining time and space in the process," I say.

“Even he can't defy gravity. You see that roll hanging over his belt? He's one fat dude, man."

"And getting fatter by the minute." The front page photo shows a larger Putz than the dumpy Role Model that started the day. Does his girth change in proportion to his stature? I wonder.

Meanwhile Melissa's radio continues to blare

One Soap at a Time

and it's driving me crazy one tawdry episode after the other what I need is my remote control Reality Converter to change the perceptual field surrounding me the radio the TV the computer screen even the formerly private Viewing Room behind my closed-eyes reality/dream dream/reality illusion/reality reality/illusion sustained delusion Rimbaud's derangement of the senses reduced to media bombardment: the medium is the message said McLuhan then now the medium *is* no prisoners allowed outside the bounds of perception all doors sealed hands on/hands off get a grip man *got* a grip man time to

SWITCH

*(*KYLE *stacks boxes in the stockroom, his head nodding to the funk rhythm leaking through his Walkman headphones.* MELISSA *enters. First, her thick shag of brown hair, then her soft brown eyes and full red lips, which part as she watches the stockroom clerk work at a pace slow beyond measure.)*

Melissa: Hi, Kyle.
Kyle: Uh, hi. I didn't see ya come in.
Melissa: You weren't looking. *(Sets her hands on her hips.* KYLE *looks. The camera's stare traces her slender frame as it tapers through a maroon sweater that hugs her small, shapely breasts and a pair of jeans caressing*

her from hips and buns down to the calves covered by brown vinyl boots with spike heels.)

Kyle: I'm lookin' now.
Melissa: Like what you see?
Kyle: Oh yeah.
Melissa: Like to see more?
Kyle: Oh yeah.
Melissa: More than you saw the last time?
Kyle: Oh yeah.
Melissa: Then, you better show me more than you showed me the last time.
Kyle: Oh yeah!

(Pushes the door shut behind her. Turns the lock. Soft cooing noises. Her boot heels rise off the concrete floor.)

I never knew Kyle had it in him. Nor Melissa. At least she'll put her fake nails to use raking his back however far they go. I'm tempted to watch, but not before issuing the following disclaimer:

THE ACTIONS OF THE CHARACTERS IN THIS WORK DO NOT NECESSARILY REFLECT THE VIEWS, OPINIONS OR ATTITUDES OF THE AUTHOR. THEY REFLECT THE VIEWS, OPINIONS AND ATTITUDES OF THE CHARACTERS WHO LIVE IN THE WORLD OF *COMMERCIAL FICTION* IN ITS MANY MANIFESTATIONS. THE AUTHOR CANNOT ACCEPT RESPONSIBILITY FOR A POLITICALLY INCORRECT PORTRAYAL OF A CHARACTER WHOSE POLITICALLY INCORRECT ACTIVITIES, INCLUDING BUT NOT LIMITED TO INDISCREET WANTON SEX, ARE INTENDED TO STIMULATE SALES OF BOOKS AND OTHER COMMERCIAL PRODUCTS THROUGH THE VICARIOUS STIMULATION OF THE SEXUAL APPETITES OF ITS INTENDED AUDIENCE.

Now that I've covered my ass let's tune back in for some nice juicy action.

[] [] []

(Close-up: MELISSA*'s firm, tight buttocks in their form-fitting denim pouch. Female Voiceover:)*

You too can have buns like these. Firm up those sagging cheeks with the

Fountain of Youth Fitness Trainer

Just five minutes a day using the Fountain of Youth Fitness Trainer and you too can turn your drooping derriere into a guaranteed head-turner.

Melissa: *(voiceover)* Want to see more?
Kyle: *(voiceover)* Oh yeah!
Melissa: *(voiceover)* Maybe next time.

[] [] []

You mean, I issued a disclaimer for *that!* If I'd known a steamy closet sex scene would deteriorate into a daytime TV self-improvement commercial I never would've said anything. I've got to admit, though, the coopting machine is getting increasingly sophisticated. I used to fast forward through commercials but now they use story elements to hook you into watching till they slam you with the sales pitch.

While I'm scratching my head on this one Kyle's running out the door nearly slams into me gallops after Melissa like a floppy hound in heat."Where can I get one o' them kits?"he shouts. I picture him stretching with the bars springs and bungee cords expecting to get at her buns through dedicated exercise. He's too spaced out to realize he'll only tighten up his own cheeks, invisible behind the baggy jeans except for the butt crack a half-inch above his beltline.

Crass commercialization. Rather than fight I'll

SWITCH

(A door with a generic "Board Room" sign. Voices murmur behind it. The camera enters a room as well-appointed as a city council chamber in an affluent suburb. Around the large oval table sit the SHELTER DIRECTOR, RALPH PUTZ, VANESSA *and a half-dozen* BOARD MEMBERS, *four of them men.)*

Vanessa: ...and so I was fired.

*(*PUTZ *and the* BOARD *turn accusing stares toward the* DIRECTOR.*)*

Putz: These are serious allegations, Emma.

Director: They're fictitious allegations. The kind you only see in soap operas.

Putz: Then, you're saying they're not true?

Director: I'm saying they're bold-faced lies.

Putz: We've had other complaints. The *GIMME SHELTER* game...

Director: *GIMME SHELTER* kept the homeless numbers high. Without those turnaway figures, we couldn't get the money to pay for the beds that we do have.

Putz: That still doesn't answer the question, Emma.

Director: I told you, Ralph—

Putz: —*Mr.* Putz.

Director: Mr. Putz. I told you, Mr. Putz, they're not true.

Putz: We have to consider the position this puts us in *vis a vis* our funding source.

Director: You should see the position *she* was in with one of the staff from our funding source.

Vanessa: That's not true. I would never do *nothin'* like that.

Putz: Like what?

Vanessa: I don't even like to talk about no stuff like that.

Putz: That's alright, Vanessa. A young, attractive lady like you...I can see how jealousy could start rumors.

Vanessa: Thank you, Ralph.

Putz: And given the recent changes in my personal situation, it's even more important that we avoid any appearance of impropriety. Therefore...

(Fade out. The phone rings. Fade in to split screen, VANESSA *on the left,* BIG BROTHER *on the right.)*

Mr. Brother?	
	Speaking.
This is Vanessa Caliente. I'm the new Interim Director of the Sister of Mercy Shelter.	
	What can I do for you?

(COMMISSIONER JOHNSON-BARR *sits behind his desk.)*

I've got to admit, at forty-one, I'm not as young as I used to be. There are times in the day when I feel my energy flagging, when I feel tired and listless. To combat that feeling of fatigue and restore my youthful vigor, I use

POWER!

POWER! restores my vitality. With ***POWER!*** I just fire a hand-picked executive assistant with an M.B.A., hire a foxy young file clerk to replace him and I'm good as new.

*(*BUBBLES *and* VANESSA *begin an endless parade of beautiful women who assemble around his desk.* KYLE *follows* MELISSA, *flexing his Fountain of Youth Fitness Trainer.* COMMISSIONER *kicks him in the seat of the pants, throws him out the door and returns to his chair, his grin widening with each woman who gathers around him. Baritone voiceover:)*

Use ***POWER!*** The Ultimate Aphrodisiac!

[] [] []

That's about what I'd expect from Johnson-Barr. He's probably getting paid more for his endorsements than he gets for playing Commissioner. He should apply for a patent on Management by Penis. He's big on Equal Opportunity screwing: he screws the people he likes and screws the ones he doesn't. Time to

SWITCH

(The stock set for an upscale restaurant. Piano music and hanging plants. PUTZ *sits across the table from* VANESSA.*)*

Putz: I still believe there are some things that are more important than power.

Vanessa: That's why I like you, Ralph. You're not like other men.
Putz: That's because I'm still a Role Model.
Vanessa: You're a wonderful Role Model, Ralph.
Putz: Believing in my message helps. I believe money is more important than power. With money you can *buy* power.
Vanessa: And love, too.

THIS PUBLIC SERVICE ANNOUNCEMENT HAS BEEN BROUGHT TO YOU BY THE UNITED STATES TREASURY.

[] [] []

Now the government's getting into the act. When I was in school I was taught we had a representational democracy. Now that we've moved past the postmodern era we have non-representational democracy. Still, it's the closest thing to an honest statement I've heard any branch of government utter.

One thing I can say about this commercial assault on the senses: it's asserting the values the Mind Control set wants us to internalize. Personally I don't see any real conflict between the Commissioner's pitch and Putz's. But then, I'm just a disgruntled, disoriented author who can't find his way from home to work or inside or outside of his own head through the stroboscopic maze of flickering stimuli. So why don't I step back and

You Be the Judge

Oy! Oy! Oy! Karaoke Court is now in session, Judge You presiding. All rise. Here come de judge, here come de judge. Court's in session, here come de judge.

*(*YOU *enter the courtroom, black robe flowing behind you. Eighteenth-century hairpiece is optional.)*

You: You may be seated.

(Nobody moves.)

You: I said, you may be seated.

(Nobody moves except the bailiff, who approaches the judge.)

Bailiff: With all due respect, Your Honor, perhaps if you sit down first, the others will follow.

*(*YOU *sit down, the others follow.)*

Bailiff: The first case is Slackborn vs. Johnson-Barr.
You: Commissioner Johnson-Barr, you have been charged with Sex Discrimination. How do you plead?

Johnson-Barr: Not Guilty, Your Honor.
You: The Prosecution will now state its case. Mr. Rother, if you will.

Wait a minute! The judge is addressing *me.* How did *I* get mixed up in this? I may be active as an author, but I was passive as a viewer. Now I'm an unwilling if not unwitting participant.

Author: On ______, ________ __ of this year, Kyle Slackborn, the plaintiff, entered the office of Commissioner Johnson-Barr, his employer, as part of a retinue. He was physically abused and expelled from the office because he wasn't a comely female.
You: He isn't a comely male, either.
Author: Your Honor, that is irrelevant.
You: But it is also true. And what does the Defense have to say about this?
Defense: Your Honor, may we approach the bench?
You: By all means.

*(*AUTHOR *and* DEFENSE *approach the bench.)*

Defense: Your Honor, we concede that the event did happen. The Defendant is willing to plead guilty to a lesser charge if certain conditions are met.
You: And what are those conditions?
Defense: That the Defendant be appointed a Superior Court Judge after his term as Commissioner expires.
You: And Mr. Rother, are you and your client willing to agree to this resolution of the matter?
Author: This does nothing to address the wrong done to my client, or to compensate him for his pain and personal suffering.
Defense: We'll raise his salary five dollars a week and not fire him for

bringing this matter to court.

You: Based on the evidence, that's as fair a settlement as this Court can reach.

Author: Your Honor, we haven't even *begun* to present the evidence.

You: Nor will you. Return to your seats. The Court finds the Defendant guilty of fourth-degree non-criminal mischief. After your term as Commissioner expires, you will be sentenced to a lifetime judgeship in Superior Court. The plaintiff will be awarded a five dollar weekly increase in salary as compensation for his pain and suffering.

Kyle: Five dollars! Way to go, man!

That's all the poor sap gets. And he doesn't even know any better. Meanwhile, Johnson-Barr gets set for life. This is some commentary on justice in this country.

Bailiff: The next case is Caliente vs. Arthur, Sexual Harassment in the First Degree.

WHAT! I thought that was purely imaginary. Looks to me like her beaver's certainly been busy, in fact, fiction or both. Now I'm caught in the blur between the two, unable to tell which is which. But she certainly *felt* real.

You: How do you plead, Mr. Arthur?

Author: Not Guilty, Your Honor. *(Sits down next to* BIG BROTHER, *his* pro bono *attorney, who shifts and fidgets edgily in the smoke-free courtroom.)*

You: The Prosecution will state its case.

(The COMMISSIONER, *wearing a "Judge Trainee" button, rises.)*

Commissioner: Your Honor, I will let the Plaintiff, Vanessa Caliente, tell the court of the injurious offenses perpetrated against her under the guise of public service.

Vanessa: *(rises)* Your Honor, this man, he did terrible things to me. At the time I was a secretary at the Sister of Mercy Shelter, where I am now the Interim Director. When I was the secretary, he used to take me to the big room we use for families who have no place to sleep at night. He would force me to do these terrible things I didn't want to do.

Big Brother: Did the Defendant actually use *physical* force?

Vanessa: Only when he tied me to the bed and whipped me.

Author: Your Honor, that's a lie.

You: Restrain yourself. You may continue, Ms. Caliente.

Vanessa: He used to take me into that room and lock the door. He said he was there to evaluate the performance of the shelter and if I didn't do what he said he would close the shelter for, how you say it, lack of performance.

Commissioner: Did he ever do that?

Vanessa: No, but he made the other Director fire me. He made her do things to him, too. These powerful men, they think they can get away with anything they want.

Big Brother: Your Honor, we can prove that Mr. Arthur was nowhere near the shelter at the time these acts were alleged to have taken place.

Commissioner: *We*, on the other hand, can prove that the former Director is currently waiting in a love motel for the Defendant, who arranged the, ahem, liaison under the guise of reinstating her to her former position.

Big Brother: Your Honor, the Commissioner has his own dubious history to deal with. Let me suggest that he arranged for this...this room to establish a basis for this fictional case against the innocent man charged with this grievous offense.

Commissioner: *(thumbs under his lapels to flaunt the Judge Trainee badge)* Your Honor, I have paid a great price for my past mistakes. I would never compromise my position in pubic service by perpetrating such an act of fraudulence.

You: If you make any further unwarranted allegations against this rehabilitated individual, Mr. Brother, I will have no choice but to raise this matter with the Bar Association.

Author: Looks like you're raising the bar, so to speak.

You: Any further outbursts from you, Mr. Arthur, and I will cite you for contempt.

Brother: Doesn't look good, bro.

Author: Not when they threaten to bag Big Brother.

You: You may continue your testimony, Ms. Caliente.

Vanessa: What he did, he did to me so many times. It was terrible, the way he made me feel....so cheap, like trash. *(Begins crying.* YOU *hand her a handkerchief.)*

You: There, there, now.

Brother: Your Honor, I request that the Defendant be allowed to testify on his own behalf.

Commissioner: Objection. Irrelevant to the case.

You: We will entertain, not so much the motion, as ourselves with his spurious testimony. Mr. Arthur, you may take the stand.

Brother: Mr. Arthur, will you state your occupation?

Author: Author.

Commissioner: Objection. Defendant cannot prove he supports himself through this fictive assertion.

You: Sustained. State your real occupation.

Brother: Your Honor, the Defendant's status as an author has a direct bearing on this case.

You: State your real occupation, Mr. Arthur.

Author: Bureaucrat monitoring homeless shelters.

Brother: Please explain to the court what actually happened.

Author: I would, if I could be sure of what actually happened. As an author transcribing what is in front of my senses, a part of me seems to exist outside chronological time. As a result—

Commissioner: Objection. The question of authorship has already been rejected by the court.

You: Sustained.

Vanessa: See how he lies. Just like that.

Author: She misinterpreted—

You: Or, more likely, you misled this woman to gain her trust, Mr. Arthur.

Author: I did nothing of the—

Commissioner: Then, what *did* you do? As a person who works as a bureaucrat—if bureaucrats can be said to work.

Author: As a person constantly blitzed by a media whose pace operates much faster than my own mind, I can no longer say what is real and what isn't. As a bureaucrat, I carry out policies that make even less sense and have less continuity than the media infomissiles I just mentioned. The lack of continuity doesn't just distort my sense of time, it destroys it. I can no longer differentiate between what actually happens and what I'm told has happened. Fact, fiction... I can't tell one from the other anymore. Events seem to happen concurrently, instead of consecutively, but as far as I know, I can't be in more than one place at the same time.

You: I can assure you, Mr. Arthur, that your sentencing will be consecutive, not concurrent.

Brother: Your Honor, I wish to change the Defendant's plea to Not Guilty by reason of mental incompetence.

You: We will let the televison audience serve as jury in this case. You, the televison jury, may find the Defendant:

(1) guilty.
(2) guilty by reason of mental incompetence.
(3) all of the above.

We will announce your verdict after this message.

Stress.

(thump-thump)

Everybody feels stress.

(thump-thump)
(thump-thump)

Stress from work.

(thump-thump-thump-thump)
(thump-thump-thump-thump)

Stress from play.

(thump-thump-thump-thump-thump-thump)
(thump-thump- thump-thump-thump-thump)

Stress from everyday living.

(thump-thump-thump-thump-thump-thump)
(thump-thump -thump-thump-thump-thump)
(THUMP-THUMP- THUMP-THUMP-THUMP-THUMP)

Stress from sitting in a kangaroo court KNOWING YOU DON'T HAVE A SNOWBALL'S **CHANCE IN** ***HELL OF BEAT-ING A BUM RAP.***

(THUMP-THUMP- THUMP-THUMP-THUMP-THUMP)
(THUMP-THUMP- THUMP-THUMP-THUMP-THUMP)
(THUMP-THUMP- THUMP-THUMP-THUMP-THUMP)
(THUMP-THUMP- THUMP-THUMP-THUMP-THUMP)
(THUMP-THUMP- THUMP-THUMP-THUMP-THUMP)
(THUMP-THUMP-THUMP-THUMP-THUMP-THUMP)
(THUMP-THUMP- THUMP-THUMP-THUMP-THUMP)
(THUMP-THUMP- THUMP-THUMP-THUMP-THUMP)
(THUMP-THUMP- THUMP-THUMP-THUMP-THUMP)
(THUMP-THUMP!)

(Cut to grinning AUTHOR.*)*

That's when I reach for

ARTFUL DODGER

the skin cream that not only
moisturizes skin dried to eczema
and soothes nerve endings frayed beyond
patience. ***ARTFUL DODGER***'s secret healing
ingredient allows me to become invisible during
any stressful moments I may find myself experiencing.

(Close-up. AUTHOR *begins to rub lotion onto himself, disappearing as he does. Cut to courtroom:)*

You: Where did he go?
Commissioner: He was sitting right there on the stand.
Vanessa: You believe a man who can do things like that?
You: I'm issuing an All Points Bulletin for his arrest.

(Cut to AUTHOR*:)*

When you need to escape from stress, use ***ARTFUL DODGER.*** I do.

A Life in the Day of (4)

"Where the hell you been, man?" Big Brother says, drawing out his tone like a 1950's hipster.

"You mean, you don't know?"

"No, man. I thought you'd *vanished*."

I rub my skin, testing for Artful Dodger residue. Nothing. Either Artful Dodger's free of clinging grease or a fictional invention. No point, then, in mentioning his recent role as defense attorney. "I *did* vanish. Sort of."

He looks quizzically at me through his cigar haze.

"You seen the Commissioner lately?" My way of conducting an indirect Reality Check.

"In what context?"

Even Reality Checks aren't as simple as they used to be. "I dunno. As the prime shaker and mover for Insurance City's downtrodden and homeless, as a self-proclaimed stud, as a political hack awaiting a judge's appointment, as a pseudo-celebrity endorsing abstract products, any or all of the above."

"I haven't seen him in any of those. But I did get a call from the chick you've been banging mercilessly at Sister of Mercy."

"Who's that?" Or which reality is that? I should ask. In matters of desire is thought separate from act? In what ways is what I've written real? In a world with no answers the role of the writer is to raise questions that make a person think outside the confines of the co-opting mechanism. If that's possible. My doing the Artful Dodger ad makes me wonder.

"The chick who decided to drop the sexual harassment charges against you."

"Now I know who you mean." Fantasy reality fiction fact whatever, I've had my share of vivid experiences with her, some pleasurable, some not.

"She's the shelter's new Director. Apparently she swung a deal with Putz."

"Apparently she swings a great deal."

Brother's silent laugh splits the screen of cigar smoke. His finger points at me as if to say "You're the one."

What product used that slogan? I've fast-forwarded so many commercials to my memory's Back Room I can't keep track....too much infodust covering the archive.

"He's their new Board President." Brother's long fingers slide the latest Extra toward me, freshly printed from the *Insurance City Gazette Online*. It's been decades since a paper's issued an "Extra Extra Read All About It" edition kids shouting on streetcorners hustling at a quarter a paper a dime or nickle before my time even pennies. Is this really the second one today?

SHELTER ELECTS CELEBRITY BOARD PRESIDENT

Putz to Use Power for Government Funds, Tax Shelter Capacity

The Sister of Mercy Homeless Shelter, beleaguered by continuing controversy concerning its low utilization rates and allegations of unethical practices, elected meteoric media celebrity and entrepreneur Ralph Putz President of its Board of Directors Tuesday.

"I intend to use everything in my power to turn this shelter around," said Putz, who recently signed a celebrity endorsement contract with POWER!

"I know he can do it," said Vanessa

See what I mean about what you write coming true? I push back from the desk, cross my legs, link my hands behind my head. "Instant communication takes more than an instant," I say.

Brother folds the paper, tucks it neatly inside his tight attache case, turns back to me. "Man, you sound like that McLuhan dude."

"It just takes time to write the article, no matter how quickly they can post it. I knew he was Board President an hour or two ago."

"How'd you find out?"

It doesn't sound like Big Brother knows everything however calm and in control he appears. No matter what Mind Control data he manages. "Moonlighting. You know I can't reveal my sources." A chuckle escapes my lips. Teasing Camaraderie.

"Not even if your ass depended on it?" Brother's high forehead creases over his mock scowl.

"Not even. But I can tell you something."

"What's that?"

"The endorsement. Putz is becoming a PR shill. He doesn't believe in ***POWER!*** He believes in Money."

Brother's long arm stretches. His hand grips the edge of his desk. "You mean, there's a difference between the two?"

"Whether there is or isn't, it could affect his credibility as a Role Model."

"He's grown so big so fast...You think he even gives a damn?"

"If he's a Role Model, it means he worked his way up from somewhere. Now that he's President of the Board, I think he'd want to keep his image intact."

[] [] []

No doubt some of you have wondered for some time about my characteristics and those of the other characters. How tall am I what color hair eyes beard? Smoker or non and related teeth coloring? Marital status number of kids formative life experiences etc. And wondered the same about the other characters. What deep underlying motives are propelling Putz's supermeth rise from PSA flak to...who knows where before I type THE END? Is he the millennial Jay Gatsby? Rags to riches without concealment? Why is Big Brother a weary civil servant instead of the omnipresent face onscreen whose verbal fingers dip into your mind and knot your neurons into the so-called proper order? If he's not Orwell's mind control myth personified who does he go home to? A wife and kiddies? A Cigar Store indian? Does he live in a Cigar Store after hours blowing his mind on chainsmoked gangsta blunts? And what about these women so brazenly sexual their lives hopscotch soap scripts? Does Bubbles LaFlamme have a husband two children ages five and nine? A hot red convertible or a neon blue neo-Bug? Does her real life revolve around choir practice two nights a week and chiming her pipes at Sunday morning service? Or Vanessa Caliente? What about this Latina whose Camp Vamp name conveys a negative stereotype? Is my portrayal of her politically incorrect or merely irreverent? Maybe she's happily married to Paco, a former gang member who now counsels in a drug rehab center. And Melissa? Maybe she's a gifted model waiting for the photo op that puts her over the top. Kyle? Maybe a great DJ or guitarist whose ability to make your feet move under you excuses his schlublike work habits and goofy sexual pursuit—if he really does either/or. The others I've neglected to mention? What about them?

No matter their habits off the page all of them are living *Commercial Fiction* from the N.Y. *Times* Bestseller List to the munchtime melodramas and other media formats for vicarious living. Where I sit (still a dubious question given the double (at minimum) life you should have figured out) I see them as living the realities readers and viewers aspire to—only several at a time, as if living a single one made a real difference. Their intrigues cross lines of medium and genre as they reinforce the conventional underpinnings carried over from the naturalistic novel. But now they're here, dipping in and out of the anti-narrative waters.

And where are you? What kind of life are you living, soap or cold wash? What are you bringing to the page? Click the left or right button in the mouse of your mind. Interact with the Virtual Fiction. Immerse yourself in the totality of the written moment. Maybe the questions will become clearer if not the answers.

[] [] []

(Medium shot. PUTZ *wearing a gray Armani suit .)*

Hi. My name is Ralph Putz, and I'm a Role Model. Now, I used to tell you not to pay attention to celebrity Role Models, people like Michael Jordan, Muhammad Ali, Mark McGwire and Harrison Ford, just to name a few. I used to tell you that most of you will never be celebrities like them. But that was before I became a celebrity. I worked my way up from a humble background very much like yours to a position that even sports celebrities would envy. Why would they envy me, a guy who showed up every day just to climb one tiny step up that long ladder to the top? Why? Because I have

POWER!

and they don't. They have money. But I have ***POWER!*** to make more money than they'll ever have. And the more money I have, the more ***POWER!*** I can buy. The more ***POWER!*** I have, the more money I can make. With the money ***POWER!*** gives me I can buy anything from cars to women to conglomerates. So, remember: every step you take up that ladder is a step toward ***POWER!***

[] [] []

"You want me back already? What's up?"

Big Brother's forehead creases over narrowing brown eyes. "You were right about Putz. The Commissioner just saw the ***POWER!*** commercial. He's not too happy about having to compete for his endorsement slot."

"You think it's competition? I thought ***POWER!*** was just showing endorsements from private business as well as government. Showing its range..."

Brother shakes his head slowly from side to side, an amused grin creasing its bottom quarter. "If you don't know the connection between business and government..."

"One subsidizes the other and vice-versa. But the players wear their Hat of the Day."

"The Commissioner thinks Putz is trying to take his endorsement hat from him."

I feel my lips twisting into a thoughtful expression. "So, where do I fit into this picture?"

"He asked me to assign one of my staff to look into it. I figure You're the One."

Now I remember it. "You're 'still having fun' with that old Coke Commercial?"

He laughs. "Diet and decaf these days...but, yeah."

"I can't start on it right away. Remember, I told you I was doing a talk show appearance? It's a great opportunity. The host is supposed to be Oprah Winfrey's up-and-coming competition."

"Start on it when you want. While I'm here having fun, you're still the one."

Four O'Clock Live

dead or five-second tape delay I'm scheduled to guest on *Afternoon with Alex*. If Alexandra Coldbreath's show moves at all it'll go network. In any case it's a big step up from my appearing on the college radio interviews nobody seems to listen to. I hope to network well enough to get an interview in front of her national audience. Even in the moment on the page (with the five-second tape delay of composition) I try to take the long view. They say it pays.

But *Afternoon with Alex* doesn't pay. Alexandra Coldbreath's staff expects her guests to be grateful for appearing because they're on the move on a show on the move. They encouraged me to take the long view.

The long view is all I get of Alex, standing off the set in her gray business suit, her breasts flat beneath a pastel blouse, her glossy brown hair pooling light between her shoulder blades. The runner's muscular bulge breaks the curve of her long calves. From what I see she's just a brunette counterpart to the bland blondes anchoring the evening news programs or the telemarketing networks. From here Alex projects an "I Can Do It But You Can't" aloofness. Maybe her backers think viewers will warm to it. Maybe they think viewers no longer identify with Oprah whose life past and present says "You Can do It Too." Maybe I'm wrong. Her staff assistant told me she read my excerpt from *Commercial Fiction*, found it delightful hilarious and incisive a perceptive commentary on contemporary life "the kind of material Alex devours." Maybe she's one of those rare media personalities who loves literature. Maybe she's one of the even rarer breed that loves literature too irreverent to be stuffy (read: My Product, Babe.) Maybe she's the antidote to the double life that leaves me too disoriented to tell if I'm at home or in the office: her interview will put me over the top as a witty personality whose book the viewers will chomp at the bit to buy. Naturally I'll start watching her show after my career-making appearance.

[] [] []

(Medium shot. COLDBREATH *sits on right, behind her desk.* AUTHOR *sits on sofa at left.)*

Coldbreath: Welcome to *Afternoon with Alex.* Our first guest today is Ann Author. Good afternoon, Ann.
Author: Good afternoon, Alex.
Coldbreath: Ann, you're writing a book. Tell me: is this your first?
Author: Before we go any further, um, I'd like to explain... I'm *an* author, not *Ann* Author.
Coldbreath: Well, Ann. Either you're the latest Boy Named Sue or you're a *terrible* ad for the latest development in sex change surgery.
Author: What I'm trying to—
Coldbreath: So, tell me, is it Ms. or Mr. Author?
Author: Mister.
Coldbreath: Tell me, Mr. Author. What do you write?
Author: I'm writing a book called *Commercial Fiction*.
Coldbreath: You're actually *proud* of writing potboilers! How does it feel to be a hack pandering to the lowest common denominator of public taste?
Author: *Commercial Fiction* isn't *really* commercial fiction.
Coldbreath: Don't tell me you're going to try to worm your way out of your sordid, sleazy contribution to the dumbing of American society, or your cynical effort to make an easy buck by fostering ignorance and even

illiteracy.

Author: I'm not. *Commercial Fiction* is a satire of the effect media has on us, as individuals and as a society.

Coldbreath: So, what you're telling me is that you have the effrontery to appear on this show and exploit the very media you're sneering and jeering at with every word you write. And *I* thought you were merely a hack. Forgive my naivete. You're *worse* than a hack. I can't *believe* I'm sitting here face-to-face with a backbiting slime ball. Worse, I can't believe that you expect our viewing audience to be so stupid as to actually *buy* a book that insults not only their intelligence but their good taste as well. *Really!*

Author: I'm not insulting their intelligence. My work is a somewhat experimental, post-postmodern fiction in which time is viewed as a multi-dimensional present instead of the linear way most people view it.

Coldbreath: What I hear you saying, *Ann*, is that your work is too precious, too "difficult" and too *pretentious* for the average reader—many of whom watch this show, by the way—to even *understand*.

Author: That's not what I said. That's what *you* said.

Coldbreath: You're accusing *me* of putting words in *your* mouth!

Author: You're *interpreting* my words before I can even *say* them. And your interpretation is hardly accurate.

Coldbreath: And *you* have the nerve to call *me* stupid on *my* show! Listen, you little twerp, you're lucky my staff even let you on.

Author: Your staff told me you read the book. You don't talk as if you did.

Coldbreath: I wouldn't be caught *dead* reading such insulting drivel.

Author: Your staff told me it's the kind of work you *devour.* Now I see what they really meant is that I'm the kind of *person* you devour.

Coldbreath: Now you're calling me a *man-eater.*

Author: You talk like the type who only comes out at night.

Coldbreath: Let's not get into my personal life, you creep.

Author: Believe me, Ms. *Coldblood*, I wouldn't get *near* your personal life.

Coldbreath: I hope you realize just *who* you're insulting.

Author: Baby, It's You.

Coldbreath: Why, you smug little cretin! You're lucky I don't whip you into line right here on the air.

Author: Even if I was into S & M, I wouldn't do it with *you*.

Coldbreath: I want you to know, you're *finished* in this business. And I *mean* finished.

Author: If your show represents this business, I'd *rather* be finished. I didn't come here to be your shark food.

Coldbreath: Oh, now I'm a shark. Well, Mister Ann or whatever your name is, you're *dead meat* as far as *I'm* concerned. *(Takes off her spike-heel shoe and hammers him repeatedly with it. As he raises his arms to protect his bloodied face, four of the channel's Security force grab his arms and legs and haul him offstage.)*

Five O'clock Bum's Rush

Who would have thought would make me the victim of an Amazon Head-hunter and the mass media's latest Celebrity Psycho? She didn't listen to me from the moment I hit the studio. For all intents and purposes,

Afternoon with Alex

"The Show That Listens to You"

I was dead on arrival. And I thought this show was going to be my opportunity to put out the word on *Commercial Fiction.* Instead I find myself brow-beaten just for showing up. If I didn't tear myself free those goons would have killed me in the parking lot. And they want people to *live* here?

Opportunity Awaits You in Insurance City

Insurance City Chamber of Commerce

I'm just passing through, en route to my humble home in

the nearest burb where taxes and insurance cost less than in Shakedown City, the land of Amazon head hunters over-priced parking lots Barbie Doll celebrities small-time big shots with egos as inflated as ticket prices who wants to give the bastards a cent after this debacle I'm so fed up I think I'll take a quick swing by

**MOM'S
ADULT VIDEO
AND
TOY STORE**

Leaving Insurance City?

Y'all come back with MONEY, heah?

to rent myself a cheap thrill for the night. Why not? A jerk-off job followed by a jerk-off TV interview. Might as well complete the day with another one and outdo Portnoy in the process. My purchase will support the economy of the town I live in: five bucks to some greasy-haired beady-eyed clerk who feels superior to his customers because they have needs and he doesn't. Hmmm...*Hot Celebrity Home Videos #1.* Wonder who's on it? Doesn't say. Typical porn. Doesn't matter, either. Today I'm just making a statement on

The Six O'clock News

PBC

What's this? I *thought* I had the right network. Maybe I pressed the wrong button. *Click!* Same thing. *Click!* Same thing again. *Click! Click! Click!*

They're all the same. What's going on here?

(Newroom. The co-anchors, who resemble real-life Ken and Barbie Dolls, sit at a kidney-shaped desk in front of the PBC logo.)

Anchor Ken: Good evening and welcome to the Six O'clock News. In a late-breaking story that surprised everyone and no one, Ralph Putz, the instant media celebrity and mega-entrepreneur, purchased all of television's networks and consolidated them under the Putz Broadcasting Company umbrella.

Anchor Barbie: The takeover sent shock waves throughout the entire industry, as network Presidents and CEOs found themselves unemployed.

(Cut to CORRESPONDENT KEN, *standing with the former* PRODUCER *of* Good Morning, Audience.*)*

Producer: I don't see how it could have a positive effect. It creates a programming monopoly, for one thing. For another, it reduces programming diversity. And it reduces the staff necessary to keep shows running on time.

(Cut to CORRESPONDENT BARBIE, *standing with* PUTZ.)

Putz: This is the Age of Downsizing. Lean, mean staffs are the way of the future.

(Newsroom.)

Anchor Ken: For those of you who haven't been keeping up with today's fast-breaking news, Putz began his rapid rise through the media industry this morning, when the former host of *Good Morning, Audience* terminated himself before the network could terminate his contract.

(A replay of the HOST*'s suicide, in tantalizing slow motion, with emphasis on the gore flying out of his brain to splatter the surrounding area.)*

Anchor Barbie: In his rise from Role Model to TV Monopolist, Ralph Putz has managed not only to move at unprecedented speed, but to break rules of time and space, as well as life and death.

(CORRESPONDENT BARBIE *with* PUTZ.)

Putz: Health Insurance is one of the items in need of greatest reduction. If there's one thing I've learned in a lifetime of being practical, it's that dead men don't need health benefits.

Correspondent Barbie: Are you saying dead women won't get hired because they need health benefits?
Putz: I'm an Equal Opportunity Employer, I don't discriminate on the basis of gender. No dead women have applied for the few openings we have. But they won't need health insurance, either.

(Cut to CORRESPONDENT KEN, *standing with the former* HOST *of* Good Morning, Audience, *whose* TROPHY GIRLFRIEND *dangles on his left shoulder.)*

Correspondent Ken: Did you ever in your life expect to be doing this?
Host: Not even in my after-life, Ken.
Correspondent Ken: There's one question I'm sure all of our televison audience would like to know the answer to: How did you manage to do it?
Host: Well, this morning I couldn't say I had much of a life. But my farewell show made me the biggest buzz in commercials. That's how I met Mona. *(Hugs* TROPHY GIRLFRIEND, *who wears a flimsy negligee over tiger-striped thong undies.)* If I didn't get a new lease on life, I certainly got one on death.

(Newsroom.)

Anchor Ken: You've seen her at work here in Insurance City.

(Clip of ALEXANDRA COLDBREATH *berating* AUTHOR:)

I can't believe I'm sitting here face-to-face with a backbiting slime ball. Worse, I can't believe that you expect our viewing audience to be so stupid as to actually *buy* a book that insults not only their intelligence but their good taste as well. *Really!*

Wonderful! they don't even show me answering her back. She makes her reputation at *my* expense.

Anchor Ken: Next you'll be seeing her nationally.

*(*COLDBREATH *berates* AUTHOR:)

...whatever your name is, you're *dead meat* as far as *I'm* concerned. *(Takes off a high-heel shoe and hammers him repeatedly with its heel. As he raises his arms to protect his bloodied face, four of the channel's Security force grab his arms and legs and haul him offstage.*)

This is what's going to establish my reputation! I went there as an author, not a human punching bag. What am I? The Mr. Bill of the Millennium? Pick me up, throw me down, run me over with a truck? Start *Saturday Night Live* and finish Sunday Morning Dead?

Anchor Barbie: That's right. Alexandra Coldbreath, the dynamic host of *Afternoon With Alex* is going network as early as this evening. PBC has announced that her show will appear every evening on all of its stations under its new title, *Alex!*

Anchor Ken: I'd say this is a real coup for her, wouldn't you, Barbie?

Anchor Barbie: I would, Ken, but not everyone agrees. Some former industry professionals think the following news has more bearing on the programming decision than her reputation as a talk show host.

Anchor Ken: Earlier today, Ralph Putz announced his engagement to Alexandra Coldbreath.

(Cut to CORRESPONDENT BARBIE, *with* PUTZ *and* COLDBREATH.*)*

Correspondent Barbie: Some former industry people think that this is merely an Engagement of Convenience. Is there any truth to these rumors?
Coldbreath: None whatsoever. Ralph has been one of my dearest friends since ten o'clock this morning.
Putz: It was love at first sight. I love the way she walks, the way she talks, and the way the ratings jump the minute her show comes on.
Correspondent Barbie: What you've just said seems to confirm the rumors.
Putz: Not if you want to keep your job.
Correspondent Barbie: You're right. I apologize for my rude behavior.
Putz: You see, I take what they call a holistic view. Money is love, and love is money. When you question your motives, not only does your relationship break down but the finances break down too.

(Newsroom.)

Anchor Barbie: And that's the way love is today.
Anchor Ken: But it does raise an interesting question. If love is money and money is love, can you live on love alone?

*(*KYLE *and* MELISSA *with* CORRESPONDENT KEN.*)*

Kyle: We're just gonna hafta. The two of us, we got fired today.

Fired! What did those poor schmucks do? They're no more inept than anybody else.

Melissa: You'll just have to keep payin' me in that ad.

Jesus Christ! I thought that was *fiction!* Too many things I'm writing are coming true. *Damn!*

Anchor Barbie: Find out more about this star-crossed young couple tonight on *Inside Report.* Meanwhile, here's another opinion.

*(*CORRESPONDENT KEN *with former* PRODUCER *of* Good Morning, Audience and HOST'S WIDOW. *The couple looks as though they haven't bathed in several weeks.)*

Producer: Do we have a choice? We can't get a shelter bed, the bums have eaten the dumpsters empty...
Widow: It's disgusting, what happens when you try to live right.

(Newsroom.)

Anchor Ken: And that's the News at Six.
Anchor Barbie: Stay tuned for *Inside Report.*

(Closing clips show Alexandra COLDBREATH *pummeling* AUTHOR *with her shoe and station Security throwing him into the parking lot.)*

Inside Report

(FEMALE VOICEOVER *comments on the action as it appears on camera.* KYLE *and* MELISSA *make out on a stack of photocopy paper boxes.)*

He was a mail room clerk, she was a secretary. Their love cost them their jobs. Will it cost them their futures, too?

(Sister of Mercy Shelter. Clips of the GIMME SHELTER *game.)*

It was supposed to be a game, but it became a sex scandal. The problem was supposed to be straightened out. Has it just gotten worse?

*(*VANESSA *and* PACO *deal glassine envelopes and automatic weapons through the shelter's rear door. Cut to* ROXIE, *a henna-haired, spandexed, overweight, gum-snapping demographic stereotype of the typical tabloid audience. She continues:)*

We'll have these stories and more on *Inside Report.*

[] [] []

(*A pair of long, shapely legs barely covered by a miniskirt curves into a lipstick red sedan. The door closes. Looking in the rearview mirror,* BUBBLES LAFLAMME *applies lipstick, rouge and mascara. As she starts to work on her eyelashes, the mirror's reflection captures a sports utility vehicle ramming her Acura's rear bumper. A chase scene ensues during* BUBBLES' *voiceover.*)

People think Blondes have it made. We get all the guys, have all the fun. Let me tell you, it's not all that easy. As you can see, I've got men who can't think of anything but slamming me from behind.

(The facade of Superior Court. BUBBLES' *sedan swerves up the courthouse steps. The SUV following her crashes into a water fountain.* BUBBLES *slinks out, parades up the steps and down a hall to the stock upscale restaurant set used for the previous lunch scenes. Cut to* BUBBLES *sitting across the table from* COMMISSIONER JOHNSON-BARR.*)*

Commissioner:...I need someone who can work with me, not only closely, but intimately as well.
Bubbles: Keep your zipper locked, kiddo. You're not a judge yet.

(BUBBLES' *voiceover:*)

As you can see, I get a lot of offers most women wouldn't refuse. But I'm different. I worked my way up the ladder, from a file clerk to a Superior Court judge. Because I'm blonde and attractive, I'm supposed to be a ditz. And that makes it a real challenge for me to dispense *BLONDE JUSTICE.*

Commissioner: I know a very nice motel where we could discuss this further.
Bubbles: Why don't you just think about it? Over coffee. *(Yanks her end of*

the tablecloth. A cup of coffee splashes into the COMMISSIONER*'s lap. Steam rises in front of his teeth-clenched grimace.* BUBBLES *stands, clicks out of the restaurant on her high heels. The* COMMISSIONER*'s agonized howl follows her all the way to the courtroom, where she appears in her judge's robe. Male voiceover:)*

Blonde Justice. Tonight on PBC. Be there.

[] [] []

Another star of *Commercial Fiction* moves upscale. A scene borrowed from *The Bald and the Ruthless* makes its way into a prime time series. Bubbles has advanced from a Soap Opera Sex Object to a Drama Star Sex Object. It doesn't look as though she slept her way to the top, either, although I can't dismiss her soapy seductions entirely; they give me too much lurid satisfaction. Real or not, the continuity created by my perceiving her in different shows endows her character with greater depth. Which makes her more real to me.

But how real is real? I'm not at work—my place of gainful employment, that is—but I'm not sure if I'm staring into a computer monitor or a televison screen. The reality of living inside my own head makes answering questions about the reality outside my head more difficult. Ads, shows, billboards, whatever...they've bopped my head so often I see their after-images in the flashbulb explosions of concussion. Lights flash, thighs flash, news bulletins flash, scenes play to resolution or commercial in a jump-cut delirium faster than crystal meth can fuel. All I know for sure is that I'm here and I'm recording it...on the computer's hard drive, on paper or on my internal drive of gray matter.

[] [] []

(Medium shot, ROXIE.*)*

Roxie: Kyle Slackborn and Melissa Dahl were in love. So in love they just couldn't keep from showing it.

*(*KYLE *and* MELISSA *in the mail room.)*

Kyle: I haven't seen you since coffee break.
Melissa: It's been so long.
Kyle: *Too* long.
Melissa: Ya mean, you're not tired or nothin'?
Kyle: I'm never too tired for you.
Melissa: Wanna take an early lunch?
Kyle: Do our double-decker sandwich? *Awright!*

(They make love on a pile of cartons, only their faces visible.)

Melissa: Oh, honey!
Kyle: Yes, baby, yes!

(The increasing passion of their lovemaking causes the cartons to pound the wall behind them.

Melissa: Oh! (*Thump.)*
Kyle: Oh, baby! *(Bang.)*

(While they moan and thrash in unison, the cartons behind them chip the plaster board wall. Dust snows over their frolicking bodies. The bumping cartons punch a hole in the wall. The hole grows larger. The cartons burst through it, KYLE *and* MELISSA *riding them as they land on top of* COMMISSIONER JOHNSON-BARR *and* BUBBLES, in flagrante delicto *on*

the convertible sofa in the Commissioner's office.

Commissioner: What the *(BLEEP!)* are you two doing in *my (BLEEP!)*ing office?
Kyle: We just kinda...dropped in.
Commissioner: Well, you can just drop by Human Resources and pick up your pink slip. You're *fired.*
Melissa: We didn't do—
Commissioner: I didn't say *you're* fired. A sweet young thing like you.... Why don't you come join us?
Bubbles: Wait a minute. I thought we—
Commissioner: We have to be flexible about our staffing patterns. Listen, sweetcakes, why don't you join in the fun?
Melissa: What do you think I am?
Commissioner: A former employee. Now, get out of here. Both of you.

(Cut to ROXIE.)

Roxie: The course of True Love never runs smooth. *Inside Report* decided to ask Commissioner Johnson-Barr if he would reconsider.
Commissioner: Of course not. I'm the Commissioner, and I have ***POWER!*** With ***POWER!*** I just fire a horny mail room clerk, hire a foxy young—
Roxie: You're not the Commissioner of *my* show, and you're not doing any *(BLEEP!)*ing commercials. That's my husband's job. It's in my contract.

*(*COMMISSIONER*'s office.* RON, *Roxie's husband, steps over, around and between the two couples writhing over the boxes, the sofa and the rug, oblivious to his workmanlike presence.* RON *wears blue jeans and a T-shirt that says* Ron's Construction Company *on its bulging front.)*

Ron: The walls too thin for ya? Tired of hearin' everything they say on the other side? Tired of party-crashers knockin' down the walls? At Ron's Construction Company we use the finest material available. We don't use no fiber board, only real plywood. Next time you wanna build your house, your apartment building, your condo or office complex, call me. At Ron's

YOU PAY TO PLAY!

(Studio. ROXIE.)

Roxie: We couldn't get nowhere with that big jerk of a Commissioner. Kyle and Melissa, you just stay the course. Don't ever forget Romeo and Juliet, the way they worked things out in the end. *(Winks and smiles.)*

Y'know, sometimes things don't work out the way we'd like them to. The Sister of Mercy Shelter put a roof over th' head of Insurance City's homeless for ten, maybe fifteen years before it had to tighten its belt and turn people away. It just couldn't make ends meet, not even with the support and cooperation it got from the government.

(The AUTHOR *and* VANESSA *spring off the double bed, cross their bodies with their arms and scamper around the floor, snatching up clothing and trying to pull it on with little success.)*

What the fuck! That one didn't even *happen.* (As far as I can tell, anyway.) It was a fantasy. Something to keep my mind occupied during a slow day at work. Vanessa was always tempting, I admit, but her ulterior motives were too obvious for me to risk my rice bowl over her. Stick with the facts, Roxie.

Roxie: The fact is, there just weren't enough beds to go around. So, Emma Frumpinger, the shelter's director, started a game called *GIMME SHELTER* to make the choice of who got a bed a little more fun, even if the loser's night in a dumpster wouldn't be much fun.

(Clips of the GIMME SHELTER *game.)*

Sometimes, y'know, games can get out of hand, like a coupla years back when Ron an' me threw our keys into— well, Emma Frumpinger decided to sweeten the pot just a little *too* much.

*(*GIMME SHELTER *game.)*

Emma Frumpinger: And the winner is...Contestants #2 and 3. You get to join me in a *menage a trois*.

And it didn't stop there. Frumpinger's secretary paid a high price for turning down her boss's advances.

(Board of Directors meeting.)

Vanessa: ...and so I was fired.

Fortunately, Ralph Putz, a friend of the shelter since the day it started, used his position as Board President to reinstate Vanessa Caliente. It seemed to be the right move at the time. But now there's rumors goin' around Insurance City sayin' the shelter's become the home base for a drug and gun dealing gang. The shelter hasn't answered our phone calls, but *Inside Report* has Ralph Putz on the line. Ralph, you've heard the rumors. What're you gonna do?
Putz: I'll have my staff look into it, Roxie.

Thank you, Ralph. And now *Inside Report* will talk with

---

better save this before I pick up no time well I'll get right back to it "Hello?"

Roxie: So, what do *you* plan to do about the shelter?

How did she even know where to call me Jesus this is getting to be too much not only my unlisted but that shadow image on the screen like some criminal in the Witness Protection Program what is how can they the invasion the violation of it all this tabloid nonsense internalizing propaganda from sheer overload what can I say except "I'll look into it"

Roxie: Don't you think there's a little conflict of interest, there?

(Repeat: AUTHOR *and* VANESSA *spring naked off the double bed.)*

"Not if Ralph Putz doesn't have a similar conflict of interest." There. That should shut her up.

Roxie: Mr. Putz is a very powerful man.
Author: Is that why you're not showing the film?
Roxie: Sweetie, I got somethin' even better to show, or my name ain't Roxie.

I don't *believe* it. She just hung *up* on me.

Roxie: So, that's the situation at the Sister of Mercy Shelter. Until Ralph Putz and the government can take care of it, you'll have to find somewhere else to sleep. And I know who can tell you where.

(Cut to RON, *wearing a T-shirt that reads* "RON'S DUMPSTER WORLD."*)*

Got no place to sleep? The shelter's full? Well, look no further. Here at Ron's Dumpster World we got places you can sleep throughout the greater Insurance City area. Wherever Ron's Construction is building, there are dumpsters to suit your every sleeping need. If you like rags and blankets, we got it. If you like good, hard plywood for your achin' back, we got that too. You want rats, maggots, cockroaches, take your pick. Here at Ron's Dumpster World we fill your dumpster sleeping needs for only a dollar a night. Meals are three bucks extra. So, when you're scrapin' the bottom of the barrel, don't forget to come to Ron's Dumpster World.

(Cut to ROXIE.*)*

A few minutes ago, we spoke to Ralph Putz about the problems at the Sister of Mercy Shelter in Insurance city, where Putz gives something back to the community by serving as President of its Board of Directors. As anybody who watches TV knows, Putz has risen from his impoverished Insurance City childhood to instant mega-celebrity and one of the world's wealthiest businessmen. Well, it seems that Putz has given something else to the hottest of TV's hottest starlets. Earlier today, *Inside Report* discovered the Model Role Model was modeling something else in an adult video called *Hot Celebrity Home Videos #1.* Here's one of the few scenes our network censors will allow us to show. We recommend that your children leave the room until the clip is over.

*(*PUTZ*'s beefy butt rises up and down as he lies on top of* PAMELA ANDALE. ANDALE *purrs and moans. Her big* blonde *hair tosses from side to side. Her legs pedal the air.)*

There's something *deja vu* about this...

Really somethin', huh? Well, *Inside Report* got an interview with Putzie boy earlier today and here's what he said:

*(*PUTZ*'s office. An imitation teak desk made from fiberboard with a "Ralph Putz, Celebrity/President/CEO" nameplate displayed prominently.* PUTZ *sits behind the desk, swiveling his black imitation leather chair from side to side as he speaks. He wears a gray three-piece suit and blue tie.)*

Roxie: So. What's up with this porn stuff?
Putz: It's not porn, it's *adult* material.
Roxie: A rose by any other name...only it don't smell like no rose to me.
Putz: It was an opportunity to expand my talents into acting.
Roxie: You didn't look like *you* were actin' there, hunka hunka.
Putz: I'm still developing my talents. But when you have an opportunity to work with an actress like Pamela Andale...
Roxie: I'd say she was doing a *great* job of acting. If they gave Oscars for X-rated movies, she'd win one hands down. Or is that legs up? It's about the only one she'll ever win, anyhow.
Putz: Pamela Andale just *happens* to be one of the great actresses of our time.

(Studio. ROXIE.)

Inside Report talked to Pamela Andale about her role in *Hot Celebrity Home Videos #1.* Here's what she said.

(Interior. ANDALE*'s Malibu mansion.* ANDALE sits by her pool in a lowcut leopardskin outfit.)

Andale: Like, you know, I consider it one of my greatest performances *ever.*
Roxie: Pam, honey. What do you think of Ralph Putz as an actor?
Andale: Well, Roxie, as an actor he has a real *gift* for paying well.

(Studio. ROXIE.)

Inside Report had a few other questions for this Putz as well.

*(*PUTZ *and* ROXIE *in his office.)*

Roxie: This kind of...*acting*. It's gonna raise more eyebrows than putzes. Have you thought about what reporters would ask you if you decide to run for office?
Putz: I won't make the mistake Bill Clinton did when they asked him whether he smoked pot in the sixties. I'll say it right now, Roxie, when I was doing my scenes with Pamela Andale, I definitely inhaled.
Roxie: How do you think your fiancé, Alexandra Coldbreath, will react when she finds out you, um, inhaled?
Putz: Well, Roxie, success has come to me so quickly, I've got to admit it has its dizzying moments. But the truth is, I filmed this movie hours before I met Alexandra, and I was engaged to her only minutes before it was released. I'm sure she understands that I in no way cheated on her. I've said with Alex it was love at first sight. And as I've said in other times and

other places, timing may not be everything but it's very important.
Roxie: So, you don't see how any of this could be damaging to your Role Model Makes Good image?
Putz: No. I don't.
Roxie: Then, mister, you *are* a real putz.
*(*PUTZ *bends forward, taps the "Ralph Putz, Celebrity/President/CEO" nameplate, settles back and grins.)*

(Studio. ROXIE, *reading a note, brushes tears away. Turns grimly to camera.)*

And so, this is my last show. I did my best to stand up for alla you Little People out there, who stand tall when they go to work or watch soaps. And I wanna thank you for bein' there for me all those nights when I needed an audience so I could keep my show. Well, tonight it looks like at least one of the Big People out there don't like my show. So, to all youse, I just wanna say, thanks for tuning in and it's been real. And to that hairy-*(BLEEP!)*ed big muckety-muck Ralph Putz I wanna say: YER A *(BLEEP!)*IN' *(BLEEP!)*HOLE AND A CREEP TO BOOT. THE ONLY TALENT YOU GOT THAT THAT BIG-KNOCKERED BLONDE BIMBO LIKES IS YOUR MONEY THRUST, YOU BIG JERK. AND I HOPE THAT ALEXANDRIA COLDFISH OR WHATEVER HER NAME IS DOES A BOBBIT JOB ON YOUR PATHETIC LITTLE WEENIE, YOU OVERSTUFFED GOLDFISH IN A HIGH-RISE TANK. I HOPE THAT HOMELESS SHELTER TURNS OUT TO BE A CRACK DEN. I HOPE THAT PHONEY ROLE MODEL—

*(*RON *rushes onstage, a storm of papers stamped "JOB CANCELLED" whirling around him.)*

Ron: Roxie. You gotta cool it. I'm almost outa business.

I HOPE THAT PHONEY ROLE MODEL CRAP YOU PULL GETS AS EXPOSED

AS THAT SHRIVELED LITTLE APPETIZER HOT DOG YOU THINK IS SO GODDAMN WONDERFUL. YOU ARE ONE SICK, TWISTED FREAK, GUY, AND I HOPE EVERY DOLLAR YOU GRAB FROM ALL THE LITTLE PEOPLE LIKE ME WHO GET THEIR SHOWS CANCELLED COSTS YOU TEN MILLION DOLLARS IN SUFFERING, YOU SELF-IMPORTANT, EGO-BLOATED SCUMBAG.

(Network Security drags ROXIE *offstage. She kicks several in the groin, gouges others in the eye.* RON *follows sheepishly.)*

[] [] []

*(*PUTZ, *jacket and vest removed, white shirt unbuttoned at the collar, stands in front of his desk, face boiling, weasel eyes blazing through his coke-bottle lenses. He rolls up his shirtsleeves, then yanks his tie to one side, a Tough Guy getting down to business.)*

This is Ralph Putz and I'm *still* a Role Model, no matter what anybody says, no matter what you see. No matter what your station in life—and I'm lucky enough to own them all, at least in this country—there's always someone out to make you smaller than you are. Especially when you're like me, a Role Model Who Makes Good. Believe me, Pamela Andale didn't think I was smaller than I am, and I sure Made Good with her. If I didn't, what kind of Role Model would I be for young people who dream about dating Sex Symbols? Some people, mostly former employees, seem to think I'm setting a bad example. (*Sing-songs)* I don't *think* so. Everything I've accomplished since this morning I've accomplished simply by Showing Up. That's what I've always told you: Show Up. *(Jabs his finger at the screen.)* When you show up, you create opportunities for yourself. *(Jabs again.)* And you show up again and again to capitalize on those opportunities. *(Jabs again and again.)* But a lot of people get bent out of shape when you do better than they expect you to. What did I do wrong? I saved *Good Morning, Audience* from cancellation, I saved television from falling apart due to the sprawl of cable TV networks. I started out with nothing, a volunteer Role Model at a homeless shelter, where I'm now the President of the Board. In a single day, I've outdone almost all Sports Heroes and movie stars combined. I've got more women chasing after me than that Roxie can shake her fat butt at. So, remember: when you succeed at Showing Up, there's always someone out there waiting to Show You Up. Show *Them* Up instead.

A Life in the Day of (4)

I had no idea *Hot Celebrity Home Videos #1* would stir such a ruckus. A lusty intermission before going back to work on *Commercial Fiction* or going out for the evening, whichever I decide to do—that was all I intended it to be. Cheap Thrills for the Single Guy. But it's cost the Tube World its most tacky representative of Middle America, the kind of people who support *Commercial Fiction* in all its guises...soap operas sitcoms dramas newscasts billboards commercials...and the all-engulfing Tabloid Reality they create. Without my seeing it the tape's given me a new take on Ralph Putz: an exemplar not so much of the banality of evil as the evil of banality. Now I *have* to watch it, if only to see Pamela Andale in action.

Hot Celebrity Home Videos #1 opens with Andale playing a guest on *Good Morning, Audience.* She's wearing a low-cut skintight see-through tanktop and ultra-micro-mini giggling at Putz's flat remarks tapping his thigh clasping his hand a great display of amusement.

The scene changes to Commissioner Johnson-Barr tying Bubbles LaFlamme to the sofa-bed in his office then to Kyle and Melissa ducking into the mailroom for a quickie. Once again this triggers my unsettling sense of *deja vu* not to mention the question of how these non-celebrities fit their skimpily-clad bodies into an even skimpier plot.

Back to Putz and Andale after the show. Putz tells her he really enjoyed her performance. She says she saves her best performances for after the show peels off her see-through outfit rubs her breasts around his hard cock then sucks it before sliding into the waterbed to receive his big-assed romp.

Wait a minute! Didn't I see this on *The Bald and the Ruthless?* Maybe I even stumbled into it. I can't tell whether it was real or imagined. If you define reality as events that result in consequences then it wasn't real.

But there I am on video stumbling in on the double-backed demon being filmed *in flagrante delicto,* too busy watching to say my lines about the Commissioner from *The Bald and the Ruthless.* Different time, different reality? Andale's lithe torso writhes under the beef-bellied humper. Her legs alternately caress him and flail the air. Her strands of hair splash sun across the waterbed's mattress. Her facial expressions shift from a dramatic pucker to a scream as her soundtrack in process crescendos whispered coos low moans gasping exhortations throttling howls glass-shattering shrieks.

It's a good thing I was beneath Roxie's notice. Or maybe not. The notoriety could have helped my writing career. Maybe it still can.

Putz rolls into beached whale position. I'm not beneath his notice. Or Pamela's.

"What are *you* doing here?" Putz is surly.

"Are you the stand-in?" Pamela Andale asks cordially.

"*Nobody* stands in for *me.*" Putz the Porn Star talking. Looks like Ron Jeremy on a Bad Body-Hair Day.

"Under the circumstances, I'd rather, uh, *lay* in."

Pamela's breezy laugh relaxes me. My eyes scan every precious curve close-up while Putz grumbles, "Nobody *lays* in for me, either."

I try to defuse the situation. "All I'm trying to do is figure out whether I'm really here or just imagining I'm here."

"I *wish* you were just imagining, pal."

"There's one way I could find out." My raised eyebrows tilt toward Pamela Andale's ample grin. Aside from epistemology I'm thinking Career Move.

"If you've got a good imagination, I couldn't help you," she says.

Obviously she has as good a command of logic as she does of her pelvic rotations. "But you *could* help me raise some interesting questions."

"Not on *my* dime," Putz huffs. His hairy gut jiggles.

A jealous porn star? What an amateur! Now I know who's bankrolling the action. Roxie was right.

"I thought I was here earlier today, too. Commissioner Johnson-Barr wanted me to stand in for him."

"Nice try, fella. But there aren't any scenes featuring him and Pam Baby."

"I was supposed to deal with the Sister of—"

"She doesn't have a sister." Putz wraps a protective arm around Pamela Andale. "And if you're really standing in for him, you'll have to go next door. He's in there filming a scene right now. But you better not interrupt him like you did me. He's not exactly a Role Model. Get my drift?"

The thumping against the wall behind them grows louder LOUDER *LOUDER* ***LOUDER***.

"The way they're going at it in there *nobody* could interrupt." Either the plaster board behind Putz and Pamela Andale is beginning to flake or I'm hyperventilating. No I'm not. The chips and dust are settling on the peak of the starlet's massive wave turning it a flat sandy color almost a natural gray.

"I'm glad to see you're learning your limitations," Putz shouts over the noise. Hardly the congenial Putz seen on morning TV.

The moans from the next room match the booming thumps. "Either they're echoing or they're having an orgy," I say.

"Wouldn't you like to know?" Putz the Celebrity Role Model lights a huge Havana cigar passes it to Pamela Andale like a post-coital cigarette. She rolls her eyes from him to me while giving it a fellatio fit for a pre-Millennial President. The jackhammering sound engulfing us doesn't seem to distract her.

"I'd like to know a lot of things, starting with what's—"

coming through the wall's a sofa bed of fucking bodies three dozen cartons of photocopy paper another pair of bodies all avalanching over Putz Pamela and me. Commissioner Johnson-Barr's body lands in sixty-nine position on Pamela but Putz has no time to grunt before Bubbles LaFlamme lands on his pole her magnificent derriere facing him Kyle Slackborn flies past us slams his head against the wall cartons crashing down on him down on me knocking me backwards onto the floor where Melissa lands straddling me her long lashes fluttering over her liquid brown gaze my hands clasp her soft but firm buns guide the ride up and down my shaft incredible the look such sweetness on her face the feel of her body so slender yet so soft who could possibly care how badly she types worth every cent right here and now my rod aching with anticipation of bursting hot come shooting into that honeysoft quim I pull her face down to mine a long twining of tongues then the explosion convulsing my body six waves of rhythm rocking me from ecstasy to a lull my half-lidded gaze meeting the grin glistening a thin coat of saliva glossing its pearl.

"Ohhhhh..." If this isn't real what is?

Muttering and grumbling around me. Clatter of boards kicked away. Boxes thud to silence.

Forget Pamela Andale and her buxom megabucks beauty. Forget Bubbles LaFlamme, her peroxide and gams. I'll take Melissa any day (or night). I tell the dreamy-eyed face floating above mine, "I never knew you felt this way about me."

Melissa's liquid gaze firms to bovine. Her little mouth gasps. "I don't. I thought you were Kyle. Where's *Kyle?* What did you do to my *boyfriend?*" Her tiny fists pound my chest at machine-gun velocity but too lightly to hurt. Then her nails razor my face. I wriggle out from under her. Stand up.

Melissa jumps up after me. I cross my forearms to block my face catch a glimpse of Kyle's body half-buried under boxes. "He's over there! He's over *there!*" I'm practically shrieking in self-defense.

She runs toward him starts pulling off the boxes. "Kyle! *Kyle!*"

A swipe of the hand wipes blood off my face. My skin feels as though it's on fire. This feels damn real to me. Especially since this event resulted in consequences. If my face isn't hacked I'm living a cartoon.

Putz is hacked Big Time. Not even a double romp Pamela Andale Bubbles LaFlamme could mellow him. "You know this guy?" He's standing buck naked next to Johnson-Barr and pointing at me.

Johnson-Barr's bouncing on Pamela Andale's belly. "Don't interrupt."

"I'm not getting paid to do this guy," Pamela Andale complains through her full-lipped pout.

"I'll take care of you," Putz assures her. "This guy crashed my party," he barks at the Commissioner of Pumps while pointing at me. "He says he's supposed to be a stand-in for you."

"Some other time, maybe. But not right now," Johnson-Barr says brusquely.

Bubbles scampers across the bed her hands knees and toes angling between the limbs of the bodies bucking on the bed. She cups Putz's cock and balls in her left hand as she rises to her knees revealing the front curve of her thighs. "I'm a big fan of your show," she says, lashes fluttering over her adoring pools of aqua. "I've admired your rise, it's so big and rapid."

"There's plenty more where that came from, baby."

A fondling of his parts as she looks him in the eye. "I know who he is." Her pile of platinum hair tilts toward me.

"I hope he's not a friend of yours," Putz says.

"He's not. He's a lech who works in my office. I can't stand the way he's always eyeballing me."

"And who do you work for?"

"I work for Commish."

"Then you'll be working for me soon enough."

Bubbles gives him a puzzled glance.

"If he's your boss, and this guy here works with you," Putz says, "then Commish is this guy's boss too. Right?"

A still puzzled Bubbles stares intently over parted lips moving as if carefully counting small numbers. "Yes,"she says. "You're right."

"I'm *always* right. Hey, Commish."

"I told you, I'm *busy*."

"Am I getting paid for this or not?"Pamela Andale is hot in more ways than one.

"I know how to take care of it," Bubbles says. She clamps her three-inch fingernails into Johnson-Barr's buns. Almost immediately the Commissioner rocks Pamela Andale with thrusts so long that he slips out prematurely spurts an arching gob splatters all over her face. "Uh oh!" Bubbles blushes.

"I'D *BETTER* GET EXTRA FOR THAT!"

"I'll take care of it," Putz says. "Hey, Commish!"

His back's feral arch relaxing, the Commissioner turns bright-toothed toward Putz.

"This guy work for you?" Putz's over-the shoulder thumb points at me.

"Big Brother sent me over here to meet with Mr., uh, with Ralph here about the Sister of Mercy Shelter," I explain. Formality seems out of place in such an intimate context.

"Your lack of professionalism is an embarrassment to my administration," the Commissioner reprimands me.

"Excuse me. I'd say we've all been caught with our pants down."

"Well, you can button up and find yourself a new job," says Johnson-Barr. "The same goes for Big Brother."

Private Dick (1)

(Replays of AUTHOR *getting fired while the Voiceover explains:)*

An innocent man uncovers a scandalous conspiracy between big business and government that costs him his job. To survive and to reclaim his good name, he starts his own business in the seamy underbelly of society as a

PRIVATE DICK

*(*AUTHOR, *in battered Stetson fedora and wrinkled beige trenchcoat, steps out of an alley's shadows into a sidestreet of brick four-storey walkups.)*

It was a bum rap.

Putz and Johnson-Barr had set up the porn scam, and I was fool enough to stumble into it. The few cheap thrills I had with Melissa, who became Johnson-Barr's executive assistant about the time I became unemployed, were hardly worth the price of my rice bowl and my reputation.

I didn't see any way to get either back. For that matter, I couldn't get the little chippie back either, not even for a one-nighter. She wouldn't return her phone calls and her new secretary said she was in a filibuster meeting. So, I bought this outfit at a thrift shop and rented this office to serve as home base for my new money-losing venture as well as a front for my living quarters: a bed embarrassingly single-sized, a refrigerator large enough for a midget's appetite and a hot plate whose burner staged wildcat strikes.

(Enters his office through a pebbled glass door with generic Private Detective lettering. Drops into the swivel chair behind his ancient oak desk, crosses his feet on his desk blotter, tips back his hat and clicks his answering machine.)

"Hi, Arthur. This is Vanessa. I'm so sorry to hear about you getting fired from your job. If you ever need any help, we can offer you part-time work running guns here at the shelter."

Just what I needed. A favor from the woman whose sexual harassment charges had started me out the door before I even knew I was going through it. The busy little beaver had made her own connections inside the clique of the rich and powerful. She wouldn't be a Sister of Mercy for long, if she ever was one. She was bucking for Mistress to Money. It was just a matter of time.

It was just a matter of time before I got my eviction notice, too. I needed money, but had no work lined up. They say things happen when you stop expecting them to. When I stopped thinking about Vanessa's message and why things weren't happening, I turned toward my nameplate.

Big Brother was sitting across from me.

2.

I couldn't decide whether a pulp or script format worked best. Since you can't make clear decisions on an empty stomach, I went with both.

Big Brother: I need your help.
Author: You didn't need it when you set me up with Putz and Johnson-Barr.

Together they sounded like a law firm. One that operated outside the law.

"I didn't know it when I sent you there," he said. "I only knew what they told me."

Big Brother didn't talk like a hipster anymore. His long face looked tired and drained. Even his black mustache sagged. He looked like a man who'd lost control of the co-opting mechanism. I wondered if he'd ever really had control of it. Or his former job. It never hurts to ask, they say.

Big Brother: I used to think I had control of it. Then things started changing. When the ads on the tube increased, the attention spans of the people sitting in front of it grew shorter. And the sound bytes on the news...It became obvious to me that there were larger forces than rhetoric at work.
Author: You take anything with you when you left?
Big Brother: I didn't. But that babe, Vanessa? After she became Shelter Director, she had a "meeting" with me. When she left, the CD-ROM with the Mind Control Data was missing from my file cabinet. In the wrong hands it can be dangerous. Her hands are just right for some things...but not this.

He'd helped me during the sexual harassment trial that took place on one of our other stages of life, real or simulated you never know for sure. I owed him a favor. And Vanessa a return call, only face to face.

We interrupt this narrative to bring you the following News Update:

The United States Government has just announced that it has sold itself to Ralph Putz. Putz announced that all legislative positions would be retained until he reviewed the situation. There are rumors of negotiations for global domination circulating in Washington, but our news sources are unable to confirm their accuracy at this time.

3.

Getting to the Sister of Mercy's door wasn't easy. The Putz Dumpster Subsidiary had circled its wagon train around the entrance and side parking lot. Ron the former owner cleaned portolets with Roxie. Processionals of shelter services professionals waved protest placards or staged somber candlelight vigils at the perimeters. The homeless people and the hardcore druggies crowded the lawn and sidewalk, fighting each other for the spot that would allow them to pound the door. Neither cunning nor brute strength would get me through them. But a twenty to Ron and Roxie got my knuckles on the door, while the other hand covered my nose.

"Arthur! I'm so glad to see you," Vanessa said from the living room as I stepped over her bodyguard, who'd fainted from the stench that followed me in. "But I won't be so glad if you don't close the door."

I closed it.

She inched so close I felt she was brushing against me, even though she wasn't. Her doe eyes rolled upward, looking big and soft. "You're here to talk business?"

"But not drug or gun-running business."

"What other business could you have?" Her breath embraced my face like morning mist, only warmer. Much warmer.

"Information."

"I got some. Ralph Putz just bought the government."

"He *what!*"

"You didn't hear the News Bulletin?"

"I'm part of the program, not the audience."

"I'm so *proud* of him. He's President of our Board, you know." She was practically hugging herself with delight.

I stepped past her, forcing myself to resist her temptress ways. You can fool me once. I don't want to find out if you can fool me twice. "That's not the information I'm looking for," I said.

"I don't got no other kind."

"Then, what's this?" I pointed to the 60" TV screen staring past her couch at us.

(Onscreen VANESSA *wears a black pullover that stops halfway down her black-stockinged thighs and pointed high heels. Her extreme beehive and makeup streaking shadows above and below her eyes make her resemble Elvira the horror-movie hostess and sex siren of* 1970's *cable TV. She sits across from* BUBBLES LaFLAMME, *the hostess for the Putz Shopping Network's "Sales on the Street" hour.* BUBBLES *wears a flesh-toned skirt and blazer. Her right leg crosses over her left, baring an impressive stretch of creamy thigh. She flashes a comely grin at the audience.)*

Bubbles: Vanessa, before the show you told me that Sister of Mercy Shelter has expanded the scope of its products and services available to Insurance City's street people.

Vanessa: Oh, yes. We recently subcontracted our overnight sleeping services to the Putz Dumpster Subsidiary. This gives homeless people a place where they can have a meal and a bed, and allows us to increase our capacity for servicing short-term guests.

Bubbles: Vanessa, could you tell our viewers more about your capacity for servicing short-term guests?

Vanessa: About *my* capacity? I don't do *nothin'* like that— not when Paco's around.

Bubbles: Then, tell us about the shelter's capacity.

Vanessa: We have many rooms available for people who need short-term shelter. We are an Equal Opportunity Shelter, which means that we do not discriminate against either gender when it comes to their servicing needs. We provide women, men or both to satisfy people who use our shelter beds for periods of four hours or less.

Bubbles: If a person like me were to come to you for short-term shelter, what could I expect to receive?
Vanessa: A hunk.
Bubbles: That sounds very satisfying.
Vanessa: We guarantee satisfaction.
Bubbles: I'm sure you do. Tell us about your other services.
Vanessa: Our new warehouse capabilities allow us to focus on the medical well-being of our service consumers. We keep more designer drugs at the shelter than you could possibly imagine. These drugs are tailored to your every need. If you want to stay up seventy-two hours and vacuum your living room non-stop, we have just the right thing for you. If you want to improve your eyesight so that you can see things that people never see even on science fiction channels, we have drugs for that too. We also have a new drug made from government surplus data that gives you concentrated doses of subliminal messages found in television shows and advertisements.

You lied," I said.

Her eyes lowered penitently, then rose in an expression at once imploring and defiant. "Don't believe everything you see on TV," she said.

Vanessa: *(onscreen)* It's our most popular product. People are trying to bang our doors down to get it.

I heard the door shake. And the floor and walls along with it. A quick glimpse at the TV told me what was going on. Outside, a crew of desperados had uprooted a telephone pole. They were using it as a battering ram to break into the Sister of Mercy Shelter. It wouldn't take long, I thought.

"I'd say I caught you with your pants down, toots."

"Arthur! Only Paco and Ralphie catch me that way."

"What about earlier today? When we were upstairs?"

"That was on TV, silly."

She had me there. When I tried to sort fact from fiction, I realized my mind had been video-surfed to insensibility. Fantasy and reality had merged into a pinwheeling kaleidoscope that changed shape every time I tried to hold it in place. Vanessa's bovine browns and delectable curves were confusing matters even further. The desperados would burst through the front door any second.

Then the back door slammed shut. I knew it had to be Paco, ready to explode with jealousy the minute he saw me.

- ***WILL THE DEPRAVED DENIZENS OF DESIGNER MIND CONTROL DRUGS BREAK DOWN THE DOORS AND TRAMPLE OUR HERO INTO A DISCOUNT DOORMAT?***
- ***WILL PACO, THE JEALOUS LOVER, SHOOT OUR HERO SO FULL OF HOLES THAT VANESSA CAN SELL HIM AS A COLANDER?***
- ***FACED WITH THIS INESCAPABLE DILEMMA, WOULD OUR HERO RATHER FIGHT THAN***

SWITCH!

Alex!

(Her face fills the screen.)

...TO ME! YOU SCUM-SUCKING RAT'S *(BLEEP!)*HOLE. TO THINK I TRUSTED YOU WITH MY FUTURE! I DIDN'T HIDE MY PAST FROM *YOU!* YOU AND THAT HOOTER-HEAVY WHORE WITH HER ORGANIC BLONDE FRIGHT WIG! AND THAT RAVEN-HAIRED VAMP AT THE SHELTER! YOU DIDN'T WANT TO BE ON *HER* BOARD. YOU WANTED TO BE IN HER *BED!* AND YOU *GOT* THERE, DIDN'T YOU? AND THAT PEROXIDE BIMBO THAT DID A BIT PART IN THAT SKIN FLICK! SHE SHOULD'VE BITTEN YOUR PART RIGHT OFF. YEAH YEAH YEAH! ALL THAT HAPPENED BEFORE WE MET, BEFORE YOU PROPOSED, BEFORE YOU SAID ALL THE *(BLEEP!)*ING MAGIC WORDS YOU DIDN'T MEAN. AND NOW YOU TELL ME IT'S NONE OF MY BUSINESS. WELL, IT *IS* MY BUSINESS. YOU COULD GIVE ME A *DISEASE!* I DIDN'T SEE *YOU* TAKING ANY PRECAUTIONS. AND I'VE SEEN EVERY LURID SICKENING DISGUSTING MOMENT, YOUR HAIRY *(BLEEP!)* AND ALL. SO LISTEN, BUSTER, I DON'T CARE IF YOU OWN THE *(BLEEP!)*ING COUNTRY OR THE (BLEEP!)ING CONTINENT OR THE WHOLE *(BLEEP!)*ING WORLD. I'M CANCELLING THIS FARCE OF A

(Putz, filing the chair across from her, presses the "Mute" button on the remote control switch in his hand. While she continues to rant and gesticulate without volume, he says:)

And I'm cancelling this show.

[] [] []

Damn! Just when I found someone who had the *cojones* to stare down Putz and his speeding monopolization of the world the man just switched her off and canceled her. Granted she was a raging mau-mau—nearly chopped off my privates a few hours ago—but it's obvious only someone as dogged as she is can stop this juggernaut. Her *Attack of the Fifty-Foot Woman* style made her the most qualified to face off against this Goliath. Me, I'm a David without a slingshot. Unless the words from my computer can shut down the media.

CLICK!

[] [] []

[] [] []

That's better.

Nothing like a little silence after a day of sensory overload. If you turn it off it can't invade you. The remote control switch gives me the same power that Putz has: I can shut off the flow of adverse information. The difference between him and me is that he controls the flow of much of the real world, as presented through the media. He's out there and I'm not. At least that's what I perceive. Who's to say it's really happening any more than Orwell's *1984* portraying wars between Oceania Europa and Eurasia that may or may not be happening except that real or fictional they're happening in the media they're happening as a perceptual presence real or not they exist as perceptions that influence the way we experience things outside us or inside. If I can't save the world maybe I can save consciousness. My own, at least.

Obviously Big Brother isn't what he used to be. He's lost Control. An unemployed Department Supervisor not an omnipresent face, like Putz.

That's *it.* Putz is becoming the new Big Brother. He's everywhere and so is his message, changing subtly then not so subtly as his influence spreads. He's still saying if I can do it you can do it too but there's a limit to what anybody can do. We can't all be Michael Jordans Michael Jacksons Mark McGwires but we *can* show up. As long as people show up they can hear the message. They show up at work they show up at school they show up in front of the tube they show up at the computer wherever the media spam

```
To: Undisclosed Recipients
From: Ralph Putz

Re: Showing up

Earn millions in your spare time. Learn success
the Easy Way. The Ralph Putz Mind Control Method
provides motivational instruction that will help
```

you become the person you want to be. This time-
tested method guarantees overnight results or your

This is too much. I'm shutting down.

[] [] []

[] [] []

There! The computer's off, the TV's off, I'm out of the studio. Sprawled on the modular sofa in the living room. Not even reading. Nothing. Nothing but watching the ceiling plaster swirl. It's fixed in place. Still those little grooves could channel stored information from one end to another link one swirl to the next pass it on a spreadsheet of the cerebral cortex in action or action painting of the cerebral cortex processing information the light the shadows in respective grooves the I/0 binaries flickering neon codes. Wonder what the information is. No. Turn off your mind. No input not even reading. The lamp. Stare at the warm glow inside the shade. Little motes of light subatomic particles protons neutrons electrons quarks neutrinos other particles discovered decades after my rudimentary study of Physics matter can neither be created nor destroyed but transformed to energy and back light one example ceiling plaster another my mind another I can't turn off I want to but I can't I want to I can't want to can't want can't want can't. What is this knocking on my door? I never have visitors. Hardly ever. I'm a solitary lives inside his own head. Better for writing. Still the world filters through whether I want it to or not. Come on. The off-switch. Click. Click!

I CAN'T STOP MY OWN MIND

so what's to stop me from opening the door unhinge the floodgates of consciousness outside me unstoppable like the ones inside me no control but consciousness itself waking sleeping live or dead I open the door.

Big Brother fills the frame. Grinning. "Hey, bro."

Talk about *deja vu!* I step back making a small sweeping gesture my left hand and forearm. He comes inside seats himself straight-backed on the half of the modular I'm not using. A nervous laugh shakes his mustache.

"I'm surprised you knew how to find me," I say.

"I still have *some* Intelligence."

"I never doubted that."

"I mean access to information."

So, what brings him here? The missing disk. No big deal to me, I say. We knew where it was before I shut down the previous narrative fact or fancy. The electronic matter of the data will transform into the energy of brain candy once the shelter's doors come down. What's the big deal? It's practically obsolete. "Even *you're* practically obsolete," I venture, hoping I don't hurt his feelings.

His forehead creases all the way to his receding hairline. "*I* might be. But the information isn't. Not yet, anyway."

Ivory Snow 99 and 44/100 percent pure? People who like people like Dial? Coca-Cola the Pause that Refreshes? Viceroy the Thinking Man's cigarette? I'd walk a mile for a mild mild Camel? L.S.M.F.T. Lucky Strike Means Fine Tobacco? "Even the Marlboro Man's running out of land."

"Those old codes, they *support* the new ones." Big Brother explains. "And the dudes banging down the shelter doors, man, they're lookin' at that as their new *kick*."

"Absorbing the old messages *we* got! Wouldn't that just turn them retro? Turn free-style dancing into the lindy? Turn punk Mohawks back into pompadours or Elvis waves?"

"That's the *surface* programming. The *core* programming is the subliminal messages, between the lines."

"What's the difference? They've got all of us programmed to one degree or another."

Brother's grin gapped wide for the first time since he stepped inside. His long finger aimed at me, then raised alongside his cheekbone. "It's the *degree*, man. A lot of those cats outside, they've become so far out of the loop the messages are losing their effect. If they swallow the data, they're back where they started. Only with Putz feeding their heads."

"And the people outside, those are the, uh, Noble Savages who can stop him?"

"As much as anybody can. It's not perfect, but..."

The air between us starts quivering. Multi-colored motes whirl out of the ceiling plaster. Like magnets they pull the light from the lamps toward them to form a hologram.

"What *is* this? I feel like I'm having a psychotic episode."

"Relax. It's"

(*the jaundiced face of* BLANCA WALL, *ace television journalist hovering above the coffee table between* AUTHOR *and* BIG BROTHER.)

Good evening and welcome to *20/200*, the news magazine for the legally blind and the short-sighted people who comprise 95% of the population, no insult intended. Tonight's show will focus on the ever-growing presence of Ralph Putz. Ralph is with us in the studio as we speak.

(Medium shot includes BLANCA WALL *and* PUTZ.*)*

Wall: Ralph, most ordinary people—celebrities, bankers, megastars and the like—find your ascendancy nothing less than astonishing.
Putz: Thank you. It's been one helluva ride so far.
Wall: But some people are inclined to say the ride hasn't been without its bumps.
Putz: Who said that? I want their names, their addresses, their job titles, their Social Security numbers, their credit card numbers...anything you have on them I want it.
Wall: You've already taken care of the people I was thinking of. Alexandra Coldbreath, for example.
Putz: A wonderful person. Unfortunately, the timing wasn't right.
Wall: What about Roxie Talkshow?
Putz: You can't have people slandering your reputation.
Wall: She's going to tell the Supreme Court you violated her First Amendment Right to Freedom of Speech.
Putz: It won't hold up in court. I own the government.
Wall: Isn't owning the government unconstitutional?
Putz: You wouldn't want to bet your job on that, would you?
Wall: Several other whimpering, sniveling, griping malcontents have also complained. Former Commissioner Johnson-Barr, for one.
Putz: Not only was he involved in a porn movie, but he failed to pay one of the co-stars. Neither of these is acceptable behavior for a high-ranking government official. He's not a Role Model.
Wall: What about you—that is, what about your relationship with Big

Brother? Wasn't he one of the ultimate Role Models?

Putz: In his time, he served his purpose well. But this is a new era, and we need new Role Models to address the needs of this new era.

Wall: Do you see yourself as the primary Role Model?

Putz: I'd be lying if I said I wasn't, and you know Ralph Putz is not a liar.

Wall: As a Role Model, what is your message to the people you role model for?

Putz: Obviously, you'll never be as successful as me, but if you do what I tell you, you can be more successful than you are.

Wall: You could be swinging on a star?

Putz: I don't want to comment on that till the News at Eleven.

Wall: We're looking forward to your announcement.

(The hologram dissolves, leaving AUTHOR *and* BIG BROTHER *in the living room.)*

"The dude is bent on world domination," Big Brother says.

"If we rescue the old data, what do we get?" This is truly freaky. There are no simple answers anymore. Maybe no workable answers either.

"What we used to have."

Another flashback to *The Prisoner:*

No. 2: You're Number Six.

No. 6: I'm not a number. I am a free man.

No. 2: You're Number Six.

No. 6: Who's Number One?

No. 2: Ralph Putz.

"That's truly frightening," I say.

"What we used to have?" Brother's hands spread palms up. "It wasn't so bad."

"It wasn't so good, either. But Putz...He stands to do some incredible damage if he isn't stopped."

"And you're the only guy who can stop him."

WHY ME?

Brother looks at me, dark eyes intent. He sits erect, arms stretched full-length, hands braced against his kneecaps. "You're a writer, man."

"So are lots of other people."

"But you're the dude working where all the subliminal programming occurs, bro."

I protest. It seems to me my work reflects the mind control that's already programmed into us even as I try to subvert it: the commercials if somewhat dated the near-antique song lyrics now used for the commercials that aren't somewhat dated the eternal retrofitting that suggests the absence of new ideas or the reluctance to use them. "How do I know you're not trying to get the data to *strengthen* Putz's as yet undeclared New World Order?"

"I got booted out of the loop, myself. You gotta trust someone, bro."

"'Trust No One,' says *The X-Files*."

"Trust No One That's In Syndication."

"Touche. Sort of." I'm not sure I buy it. But I've done other things I haven't bought. That much I'm sure of. "What do you want me to do?"

"Write. When you write you create your own programming."

"Maybe I can replace the old codes we get back with some new ones. Or eliminate codes permanently."

Brother's face lights up for the first time since he's been here. "Now you got it, bro."

"Then, let's do it."

He follows me into my studio in the back room of the house. Closes the thick oak door behind us. The minute my monitor screen turns white he clicks off the lights.

"Where do you want me to go?"

"Back to *Private Dick.* How's that?"

"Too risky for the moment. Vanessa knows I'm onto her, and Paco's coming through the back door."

"Got any other ideas?"

"Just a shot in the dark."

Blonde Justice (1)

(JUDGE LaFLAMME *climbs the courthouse steps. Even her black designer suit can't contain her swaying curves. Former Commissioner* JOHNSON-BARR, *wearing panhandler's rags, follows her.*)

Johnson-Barr: Baby, baby, don't you treat me this way. I'll be back on my feet someday.
LaFlamme: Hit the road, Jack.
Johnson-Barr: Come on, Bubbles. It wasn't like this before. I was gonna be a *judge*. Like *you*.
LaFlamme: It's different now. I'm a Role Model.

Big Brother: You sure this is the way to go?
Author: I don't think she's swallowed the *entire* Putz line. But we'll find out.

Johnson-Barr: You and I were in that movie together.
LaFlamme: Under conditions that bordered on sexual harassment.
Johnson-Barr: I didn't hear you complaining too loud.
LaFlamme: I wasn't. But then, I didn't violate Pamela Andale or the terms of her contract.

Johnson-Barr: What you're saying is, you didn't piss off Ralph Putz.
LaFlamme: That's *one* thing I didn't do to him.
Johnson-Barr: But you admit to the others.
LaFlamme: Like you, I did what I had to do to get where I wanted.
Johnson-Barr: "Bubbles LaFlamme, the Swinging Judge."

*(*JUDGE LaFLAMME *swings her purse toward the camera. At the thud of impact and the grunt of* JOHNSON-BARR, *the camera cuts to the headline:)*

FEISTY FEMME LaFLAMME FELLS COURTHOUSE FREELOADER

The justice known to political insiders as "the swinging judge" put a new spin on her reputation today as she defended her integrity with a swing of her purse on the courthouse steps earlier today. Superior Court Judge Barbara LaFlamme, responding to aspersions cast on her character by former Information Agency Commissioner Johnson-Barr, dealt him a blow to the head, which caused him to tumble down the courthouse steps.

"I don't claim to be a saint,"Judge LaFlamme said in defense of her action. "But the one thing I did that he didn't do was go to night school for eight years."

Johnson-Barr, whose extracurricular activities have already cost him a judicial appointment, grumbled, "No comment."

Big Brother: Whoo! Has she got a *temper!*
Author: She's got a backbone too.

(That's what I was looking for. There's more to her than the Fox in the Box commercials show. Her freakout rejection of me on *The Mating Game* this morning wasn't enough to go on—if it ever really happened. I can't say for sure that it did any more than I can say for sure that she was in Johnson-Barr's office during *The Bald and the Ruthless* and what she was really doing there. For that matter I may or may not have been in the cast of *Hot Celebrity Home Videos #1.* If nothing is certain in reality anymore, including reality itself, the imagination still allows me to explore other possibilities.)

(Interior. JUDGE LaFLAMME*'s chamber.* AUTHOR *and* BIG BROTHER *sit in front of her desk.)*

LaFlamme: Speaking off the record, you're quite correct. There is no statutory authority that allows him to buy the government. But from a procedural standpoint, you're very naive in thinking that a ruling can be made without a case to rule on.

*(*AUTHOR *and* BIG BROTHER *walking down courthouse steps.)*

Big Brother: And here I thought she was only a file clerk.
Author: I couldn't see beyond her looks, either. She's very astute.

I turn away from the screen. "Sounds like we have a chance with her. But we've got to lure Putz into making a move."

Brother's straightahead stare suggests that he's stymied.

"We have to get our hands on the CD-ROM," he says. "Putz probably thinks he owns it."

"What if we go back and get it from the shelter?"

"No way, man," shaking his head. "Putz's goons know who we are."

Brother's face sags around his stare. He knows the Putz Dumpster Subsidiary is the front for Putz's Security. "What about Ron, Roxie Talkshow's old man?"

"That just might work," I say, reaching to lift my Stetson fedora off its futuristic mannequin head. "But we need more people."

"Word on the street is that Kyle and Melissa are banging on the shelter door with the rest of the mind candy crowd. Putz canned them along with

Johnson-Barr. They followed us out the door right after the porn fiasco."

"I thought Melissa was his Gal Friday, not to mention several other days."

"Say *what!*"

Big Brother must not have access to the same information real or imagined that I do. "Never mind. You check with Ron. Then find Kyle and Melissa, and go to the back door of the shelter."

Brother looks as though he thinks I've popped my cork too far before New Year's. "While you stay here! Shit, man. What good does that do?"

"You've got your programming script, I've got mine."

[] [] []

Stay Tuned for tonight's Special Episode of* Blonde Justice, *in which Judge Bubbles LaFlamme and the characters from* Commercial Fiction *team up with the

Private Dick (2)

5.

"My boyfriend's home," Vanessa said. "You're in big trouble now."

"Maybe I am, maybe I'm not." Everything depended on what happened next. If my scheme didn't work, I was in bigger trouble than she thought.

The back door slammed shut a second time. Paco's stamping feet stopped in the kitchen. A trample pulled up after them.

"Hey! How *you* get in?" I heard him say.

Three familiar voices murmured in the kitchen. Just the way I'd planned.

I had them right where I wanted them. All of them.

Vanessa spun toward the kitchen, then turned back to me. "You better leave," she said. "His friends came with him."

I stood my ground. "Look at the screen, *chica*."

(Voiceover.)

And now the Virtual Court segment of *Blonde Justice.*

"How you switch my channel?"

I grinned into her incredulous face. "I'm only a Private Dick," I told her. "Maybe the network stuck this virtual segment thing in so it could compete with all the interactive courtroom shows on cable."

"I don't believe you." Her eyes didn't have the seductive look of a minute ago. Anger flashed through them. Then worry.

(Virtual court. JUDGE LaFLAMME *addresses* VANESSA *from the bench.)*

And how do you plead to charges of illegal possession and distribution of Classified Government Information?

"You take that CD-ROM, I blow you suckers away," Paco shouted in the back room.

The worry in those soft browns turned to panic. "Paco. Let them have the disk."

(Courtroom.)

Judge LaFlamme: Again. How do you plead?
Vanessa: Innocent! *Innocent!*

"Paco! Let them take it. *Paco!*" She ran into the back room. "I'm in Virtual Court behind that stuff."

"Take it. Out this way. *Andale.*"

Beautiful! I knew what would happen as soon as they hit the street. Now it was my turn to duck out.

6.

FORMER OFFICIAL, CLERKS ARRESTED FOR TREASON

Tandem Charged With Theft of Confidential Government Documents

Putz had taken the bait.

No sooner than Brother, Kyle and Melissa had ducked out the shelter's back door, somebody, probably Vanessa, had called Putz. He'd needed the data to move his career and she got it for him by moving under Big Brother. Maybe our shelter affair and the sexual harassment charges she filed against the Shelter Director and me were ways to meet Big Brother make it with him and make off with the data. She probably told Putz some counter-government agency—no doubt headed by me—had robbed the CD-ROM from her. At gunpoint. She'd say anything to save her pretty little butt, I figured. Even if LaFlamme's Virtual Court showed otherwise. Vanessa could claim either was fact or fiction. Me, I'm just a first-person narrator, working within my limits. I didn't care about the gun-running and hooking at the shelter.

Neither did Judge LaFlamme, aware as she was of political realities.

There were, of course, other realities to be aware of.

"I think my show has gotten out of hand," she told me in her chambers before the arraignment. "The Virtual Court segment wasn't something I agreed to when I spoke with your script consultant."

"He told me he needed it to make the case. I'll tell him to keep a lid on his imagination next time."

"Please do. All my years of study have taught me there's no such thing as Creative Jurisprudence."

"There's always a first time."

"Nothing will happen outside this chamber if you don't return me to

Blonde Justice (2)

(Courtoom.)

Bailiff: *Oy! Oy! Oy!* Court is now in session, Judge LaFlamme presiding. All rise. Here come de judge, here come de judge. Court's in session, here come de judge.

LaFlamme: In the future, Bailiff, I would appreciate your refraining from such politically incorrect pronouncements. You may be seated.

(Only the BAILLIFF sits.)

The rest of you may be seated, too. The Court will now hear the charges against Big Brother, Kyle Slackborn and Melissa Modell.

Melissa: That's model, Your Honor. My last name is Dahl.

LaFlamme: I stand corrected. Mr. Johnson-Barr, you may present your case.

(JOHNSON-BARR, *freshly-scrubbed and newly-appointed, steps forward.)*

Johnson-Barr: Your honor, Ralph Putz, President/CEO and owner of the United States government, asserts that the data in question was illegally taken from his place of safekeeping at the Sister of Mercy Shelter.

Vanessa: Your Honor, he's lying. That CD-ROM, it was *never* at my shelter.

LaFlamme: Where was it found?

Johnson-Barr: On the person of Kyle Slackborn, Your Honor.

LaFlamme: Was he inside the shelter when the arrest was made?

Kyle: No, I wasn't. I was walkin' down the street with my girl an' my friend. Next thing I know there's all these guys beatin' on me an' throwin' me to the ground. My girl an' my buddy, too. Good thing the cops showed up.

LaFlamme: If the CD-ROM was never at the shelter and the defendants were arrested outside the shelter with it in their possession, there is no case.

Johnson-Barr: Your Honor, the data on that disk is the property of the U.S. Government, which, in turn, is owned by Ralph Putz.

LaFlamme: According to my research on the issue, the government itself is not physical property. As such, it cannot be sold. My research further indicates, based on an analysis of presidents, senators, congressmen, interns and lobbyists, that the government does not qualify under existing statutes as intellectual property, either. As such, the data in question resides in the public domain. Mr. Putz, this Court hereby orders you to cease and desist from any further misrepresentation of yourself as owner of the government and the document in question.

Putz: Baby, baby, don't you treat me this way. I'll be back on my feet someday.

LaFlamme: Hit the road, Jack.

Johnson-Barr: Maybe we could explore the ramifications of this ground-breaking court decision over dinner?

LaFlamme: Hit the road, Jack.

*(*LaFLAMME *pounds the gavel. Background music: Ray Charles and the Raelettes continue "Hit the Road, Jack" as the courtroom empties. Cut to courthouse steps.* LaFLAMME, *out of her robes and into her miniskirt, sways down them, toward the camera detailing her every move.)*

A Life in the Day of (5)

It was over. I'd achieved closure as best one could without cessation of all realities created through the continuing process of individual perception. Perception is like time. You can speed it, you can slow it, you can establish arbitrary constructs to document or channel its flow, but you can't stop it without stopping existence itself.

Our existence wasn't stopping. It continued as a victory celebration at Ron's Sports Bar. The Sports Bar was the logical extension of the concept of the Megabook$ chain, which operated on the premise that profit mattered more than product. During the time I was (or am) writing *Commercial Fiction* the Megabook$ chain's CEO decided to replace its browse-over-coffee-then-buy-a-book approach with one more interactive: a rectangular liquor bar replaced the coffee bar at the corner of the store as well as the Customer Service Center fiction poetry mystery sci-fi biography and children's sections. Current events became the four screens hanging from the ceiling one for each side of the bar three broadcasting sporting events such as Extreme Sexual Bragging No-Holds-Barred Drink Hustling and Armchair Sports Utility Vehicle Imagination Competitions. The other screen, which Brother and I sat in front of, previewed the Eleven O'clock News featuring Alexandra Coldbreath.

"Looks like she played kissy-face with Putz," Brother said.

"She's not the only one playing kissy-face." I looked toward the booth where Kyle moonlighted as a DJ. He and Melissa had steamed the windows pretty quickly.

"What did you expect?"Brother said. "Since when do DJs play anything in sports bars?"

"They must play 'Take Me Out to the Ball Game,' at least. Or the theme from *Rocky.*"

"You're talking Premium Channels, man. The stuff that's on these screens, Premium Channels don't broadcast that. We're talking low overhead, man. *Real* low overhead."

"Anyway," I said, "it looks like papa's got himself a brand new bag."

Big Brother leaned back, looked at the booth rocking on its foundation, then flashed his grin at me. "He'll be back in the mail room tomorrow and she'll be back typing five words an hour and admiring her nails."

"Yeah..." Everyday Reality has its practical drawbacks. "But maybe she'll get more work as a model."

"You said she was a model. I didn't."

"You mean, she isn't?"

"Not according to my script. Maybe according to yours..."

So what *was* the Reality of all this? Did he imagine one thing and me another? I thought my writing to retrieve the stolen CD-ROM was as real as what was broadcast, a creative intervention that redirected reality to serve a common purpose. Maybe the Mind Control Data never existed outside our respective fictions—or in quite the same way inside them.

Bubbles LaFlamme arriving on the arm of Johnson-Barr made me doubt everything I had done. Especially when I heard her call him "Commish." While I was into *Blonde Justice* was she still into *The Bald and the Ruthless?*

"Hey, Commish!" Big Brother waved, then turned back to me.

"So, what did we really accomplish?" I asked my once and current boss.

"A work of *Commercial Fiction,*" he said. "Powerful, attractive individuals stooping to the lowest possible depth to gain the highest possible stakes."

"While leapfrogging from game show to talk show to soap opera to detective show to courtroom drama with a bunch of other shows in between, not to mention all the commercials. So, what's the point?"

"That shit happens everywhere. If it doesn't finish one place, it keeps going through the next."

"You got the data back, didn't you?"

"That's for you to decide." Brother pointed to the scoreboard directly over the cash register:

Putz	**0**	**U.S.**	**2**
Putz	**5**	**TV**	**1**
Putz	**1**	**Big Brother**	**1**

and so on down the line. Nothing was as clear-cut as the resolutions of plot suggested. Putz had lost some ground but not as much as I thought. What kind of closure would really fit into the twenty-four hour Aristotelian formula? Not much if Bubbles LaFlamme had regressed from tough-minded jurist to long-legged file clerk. Not much if Ralph Putz still controlled the airwaves as a result of his one-day rise up the media ladder. Maybe the news would tell me.

[] [] []

(Newsroom.) Good evening. This is Alexandra Coldbreath and here is

The Eleven O'clock News

She doesn't have a co-anchor or an attitude. Either she's played kissy-face with Putz or he's losing power.

Coldbreath: In an unprecedented legal decision, Justice Barbara LaFlamme ruled that the United States government does not have the authority to sell itself to celebrity entrepreneur Ralph Putz. The landmark decision brought legislators and lobbyists alike back to Capitol Hill from their vacations in Bermuda, Paris and Timbuktu.

So what's LaFlamme doing sitting at the far end of the bar flirting with her courtroom and carnal adversary? What happened to "Hit the Road, Jack?"

Coldbreath: We have Ralph Putz on location in his office tower to discuss the effect of the decision on his landmark entrepreneurial efforts. *(Cut to* PUTZ *onscreen.)* Ralph, what is the effect of the court's decision on your landmark entrepreneurial efforts?

Putz: Naturally, it's a major setback, Alex. But in the five minutes since the decision was made, I've managed to buy Europe, Russia, China and India, countries where the rules about buying governments are less stringent. Africa, South America and Australia are a little more resistant because they can't see the opportunities I'm offering them, but I'm confident that I'll achieve global domination before I start Day Two on *Good Morning, Audience.*

I turn to ask Big Brother what we've really accomplished beyond an exercise in the futility of the imagination only to find myself staring at the newscast from my cushioned sprawl on the computer chair eyes glued to the screen filling with typeface look down fingers blurred movements life of their own or drawn from some other part of me this story is it mine or not am I writing it or am I a channel for unknown forces recording the news as it happens or just before depending on how tuned in I am either way the tableaux of Big Brother next to me on one stool the newly regressed Bubbles LaFlamme on another with Commissioner Johnson-Barr continuing a daylong rendezvous through narratives soaps and prime time series that may or may not be real or any more real than my sitting here alone which may or may not be either

---

(Studio.)

Coldbreath: In a related matter, the Sister of Mercy Shelter, a recent source of controversy, appears to have righted itself. Under scrutiny for sex scandals, gang activity and other illicit activities, the service organization seems to have renewed itself with a fresh sense of purpose.

(Cut to shelter image onscreen behind COLDBREATH. VANESSA *stands in the foreground, her breath clouding in wintry air. The former* SHELTER *DIRECTOR, the former* PRODUCER *of* Good Morning, Audience *and the*

WIDOW *of the show's former* HOST *stand behind her.*)

Vanessa: The Board of Directors and I felt it was necessary to resolve these matters as soon as possible. We called an emergency meeting which just ended. We have agreed to subcontract our shelter services to the Putz Dumpster Company, and we have re-hired the former Shelter Director to serve as a link between Sister of Mercy and Mr. Putz's firm. Sister of Mercy itself has revised its mission so that it can provide much-needed gun-running services to revolutionary forces in the third world countries that have yet to be purchased by Mr. Putz. My boyfriend Paco and I will oversee this new and liberating venture.
Coldbreath: We also have representatives of the homeless population with us. Tell me, how do you think the shelter's new direction will impact your lives?
Widow: At least we have dumpsters to sleep in. The guards provide more security than my deceased husband ever did.
Producer: I look at it as an opportunity to start a second career. If I can launch the career of a man like Ralph Putz, I can use my expertise to help the shelter become one of the finest weapons suppliers to the Third World.
Coldbreath: Thank you. We'll be back after this message.

The former HOST *steps in front of his* WIDOW *and the former* PRODUCER *of* Good Morning, Audience.*)*

There's an old adage that says, "When life hands you a lemon, make lemonade." In terms of lemonade, this joke of a producer claims he launched Ralph Putz on a career whose meteoric rise surpasses all laws governing time and space. And my former wife talks about security in a dumpster! If you ask me, the two of them have turned the lemons I gave them into

Citron Frozen Lemonade

Citron Frozen Lemonade is made of the finest lemons you can find among the throwaways of the restaurant business. Its special mold guarantees to reverse the effect of any penicillin mold you may find growing on other discarded food products. What's more, Citron Frozen Lemonade is guaranteed not to thaw under any circumstances, not inside microwave ovens, not over Sterno burners. So, when life hands you a lemon, don't waste your time looking past the Dark Side. Take your lumps with Citron Frozen Lemonade:

The Frozen Lemonade that stays frozen, no matter what.

(Grins and walks into the shadows.)

Producer: I–I thought you said I had a job here.
Vanessa: It was a decision made by the CEO. I hope you understand.
Widow: What about the dumpster?
Vanessa: I'm sorry...

(Voiceover:)

Citron: a subsidiary of Gloom and Doom Specialty Products, Incorporated, Ralph Putz, President/CEO.

[] [] []

This news isn't coming to me from a TV screen. It's my computer monitor. I've spent all day here or in the office connecting scripts that collapse real time within the Aristotelian twenty-four hour criteria for drama Strindberg's expressionistic violation of it in *The Ghost Sonata* and my own quest for closure to events squeezing lifetimes into media times except the life outside my mind's control resists closure even as it adheres to the mind control programming of the government and the media their synergistic symbiosis intertwining beyond the umbilical snip of any streaming consciousness programmed reprogrammed deprogrammed. Death extends beyond life for commercial purposes time speeds up for greed and media fast breaks down the court beyond *Blonde Justice* and its remote-clicked alternatives one reality's as good as another or at least as programmed. My brain's fried. So I'll **CLICK!**

A Life in the Day of (6)

off the monitor watch the tube at the foot of the bed wind down from the illusion of real life by watching other illusions one eye barely open the volume down low murmuring in the gray cathode aura surrounds the technicolor projections an oddly soothing fluorescence seeping into my alpha waves no doubt further conditioning subliminal programming the bachelor's doze to the comfort whatever plot onscreen offers a Japanese sci-fi rerun a Ginsu knife commercial anything to rest the mind from a world where holograms exist inseparable from concrete.

[] [] []

(The former HOST *of* Good Morning, Audience *sticks out his tongue and thumbs his nose. He pulls the trigger. At the audience's collective shriek, the screen goes black. Then, his voiceover while the program title appears:)*

Good evening and welcome to

Midnight Movie Madness

*(*HOST *raises his head off the desk, grinning, a bullet hole near his temple.)*

Tonight's 1957 Japanese cult classic features one of today's most celebrated figures in his first walk-on appearance ever. Some of you might be surprised to learn that this faster-than-overnight success has been around since the days of *King Kong* and *Godzilla*, but, like the creature he plays in his big screen debut, he punches a time clock that is out of this world.

(The grinning HOST *shoots himself again. As he falls forward, the screen goes dark, then:)*

[] [] []

UNIVERSAL PARTICULAR PICTURES

[] [] []

The Putz from Planet X

[] [] []

begins with a space shot of Planet Earth. From the blackness behind it a blip of light arches through space to strike Tokyo. Cut to the maternity ward of a Japanese hospital. The camera focuses on an infant Ralph Putz, presumably born with thick glasses and shaggy mustache, wearing diapers held together with a two-foot safety pin. The English subtitles ask where did he come from? Nobody knows. He's growing so big so fast was he in the atomic blast at Hiroshima? Nobody knows. Cut to exterior shot of building.

High-pitched female screams from inside. The building explodes. Rubble flies toward the viewer like debris from an early 3-D movie.

[] [] []

High-rise office building in L.A. A Bubbles LaFlamme lookalike plays secretary to an R. Thorpe, Director of Character Development. Putz looks into the office, his nose and one bespectacled eye filling the floor-to-ceiling window. LaFlamme shrieks, recoils with horror. Glass shatters. A huge hand of baby flesh pulls her outside. R. Thorpe, a trim-haired dark-suited precursor of *Commercial Fiction's* author, rushes outside to watch his fiancé devoured curve by delicious curve.

[] [] []

(1950's newsreel. Headline:)

ENFANT TERRIBLE TERRORIZES COUNTRY

Midwest Corn Crop Crushed, Chicago Meat Business Slaughtered

[] [] []

*(*WALTER WINCHELL:*)*

Good evening Mr. and Mrs. America. In his temper tantrum across our nation the *enfant terrible* that escaped from Japan has decimated this country's cash crops meat supplies industrial facilities and everything else that crosses his trampling path. This mass of dimpled destruction has crushed the city of Chicago chewed up Cleveland pummeled all of Pittsburgh and dealt a death blow to Delaware. His latest swath against the American Grain suggests that he is advancing inexorably toward our nation's capital where he will rattle the entire republic to death. If you ask me, I wish the fat little brat would grow UUUU***UUUPPPPPP***—

(A pudgy hand pulls WINCHELL *out of his booth.)*

[] [] []

(Pentagon. Interior. Master Sergeant VANESSA, *of the Women's Army Corps, leads a disheveled and unshaven* R. THORPE *into the laboratory of Dr. B. Brother. Also in the laboratory is Major General* BARR JOHNSON.*)*

Thorpe: You've got to believe me, Dr. Brother. This Big Baby has ruined my life. I've seen it do things...things no man should ever have to see.
Dr. Brother: Our Developmental Psychology Unit is researching the matter.
Thorpe: I wish it would *grow up*.
Dr. Brother: Be careful what you wish for. You might get it.

(A wingtip shoe and gray trouser leg stamp through the Pentagon's roof, barely missing R. THORPE *and* BROTHER. *Cut to* GENERAL JOHNSON*'s office. An aerial map shows the eastern seaboard crushed by footprints.)*

Dr. Brother: According to our research, he is a mutant life form from another galaxy. We have no way of knowing how he got here or why he came. We've analyzed tissue samples and it appears to us that he grows when exposed to radioactive materials.
General Johnson: In his journey east, he has devoured all of our nuclear missiles. It is safe to say that your fiancé is not all he eats.
Thorpe: Thank you. That's very reassuring.
General Johnson: It is also safe to say that nobody, male or female, is safe from this creature's truly monstrous appetite.

(On the General's wall hangs 1950's pin-up girl, PAMELA ANDALE.*)*

[] [] []

*(*PUTZ *walks through midtown Manhattan, crushing Times Square and Greenwich Village in successive strides. En route he picks up* MELISSA *and* KYLE, *two tourists on their honeymoon, and chews them like Mounds bars. When he snatches* PAMELA ANDALE *from her on-location filming and carries her to the Empire State Building, the armed forces go into action. Fighter planes strafe* PUTZ, *who brushes them away as if they were gnats. Napalm bombs merely enrage him. He picks up trailer trucks and throws them at the bombers.)*

General Johnson: We're down to our last line of defense.
Dr. Brother: We grew this from a patriotic volunteer's tissue samples.

*(*BROTHER *pulls an immense white sheet, revealing a hundred foot high* ALEXANDRA COLDBREATH.*)*

Thorpe: Incredible! If anybody can beat him, it's her.

Uh...*(yawn)*...How did I know *that?* In 1957, no less...

---

*(*ALEXANDRA COLDBREATH, *in her gray anchorwoman suit, strides with grim determination toward the Empire State Building.)*

Dr. Brother: This is our last hope.

*(*COLDBREATH stops *at the base of the Empire State Building.* PUTZ *is too*

preoccupied with PAMELA ANDALE *to notice her. His gigantic fingers poke and fondle the terrified blonde who writhes in the palm of his hand, trying to dodge the pawing gestures that inadvertently strip away her clothing.* COLDBREATH, *standing as high as his kneecap, digs her hands into her hips.)*

Coldbreath: LISTEN TO ME, YOU SCUM-SUCKING PIECE OF SPACE BLUBBER. WHATEVER PLANET SENT YOU HERE, DON'T THINK YOU'RE GOING TO TAKE OVER THIS WORLD. WE HAVE GOOD, RIGHT-THINKING AMERICANS HERE, GOOD RIGHT-THINKING AMERICANS WHO HAVE TURNED BACK EVERY ENEMY THAT'S EVER TRIED TO TAKE US OVER. THE BRITISH EMPIRE, THE AMERICAN INDIANS, SANTA ANA AND HIS MEXICANS AT THE ALAMO, THE GERMANS AND THE JAPS. AND WE'LL BLOW THOSE COMMIE RUSSIANS TO KINGDOM COME IF THEY EVER TRY TO BOMB US. SO DON'T THINK WE'RE AFRAID OF YOU AND ANY OF YOUR CREEPY MARTIANS OR CREATURES FROM PLANET X OR WHATEVER THEY ARE. YOU'RE DEALING WITH DECENT, GOD-FEARING PEOPLE HERE AND NOTHING YOU CAN DO WILL STOP US FROM STOPPING YOU. GET THAT STRAIGHT, YOU BLUBBERY, FOUR-EYED, FAILED IMITATION OF A SCHNAUZER!

Good Old Alexandra. Some things never change.

---

*(*PUTZ *stamps up and down, crushing several blocks of office buildings. His bawling bellow breaks every remaining window in midtown Manhattan. He kicks* COLDBREATH *onto the top of Grand Central Station.* COLDBREATH *runs back to the Empire State Building, springs at him. Her teeth dig into his thigh. The combined pitches of her rabid growling and his pained howl turn Manhattan to dust as far north as Central Park. He lifts his leg and tries to*

shake her free. No luck. He raises the hand holding PAMELA ANDALE *over his head and starts a hurling gesture.*

I don't know what came over me. Some call it courage. Some call it a testosterone surge. All I know is that before I knew what I was doing I was running into the deserted street. I vaulted into a red 1955 Cadillac convertible and raced forward. Don't ask me what plan I had in mind. I was running on sheer instinct. And guts. No creature from outer space was going to destroy America's favorite pin-up girl. It was time David fought Goliath.

The creature hurled Pamela Andale at my advancing vehicle. She struck the back seat with such force that the convertible spun a full circle. Sheer centrifugal force hurled me into the back seat on top of America's favorite pin-up girl, who fixed her stunned gaze on me and said through a dreamy grin, "My hero!"

Pamela Andale planted kisses of gratitude over my face and lips while I checked her body for injuries. Her breasts were full, firm and what few bruises I found I tried to make better by kissing them. The landing hadn't damaged her thighs, which were shapely and without a trace of cellulite. I rested her ankles on my shoulders and tried to bring out my tool to measure the extent of possible internal injuries when all of a sudden I felt a stiff breeze pull my fedora off my head. America's favorite pin-up girl looked past me and screamed. I turned. There was the gigantic creature, one block-sized foot ready to crush America's favorite pin-up girl and me into the asphalt.

Somehow— don't ask me how—I did what had to be done, not only to save our planet but to save America and its favorite pin-up girl. In her moment of panic, Pamela Andale pushed me away. My foot hit the accelerator. The car lurched backward. Time moved in slow-motion. I saw every detail of the Cadillac's tail fin puncturing the creature's big toe.

He howled in pain and dropped Alexandra Coldbreath onto the hood of the car, catapulting Pamela Andale and me to safety in the cushioned ruins of a garment district warehouse.

(Upon impact, PUTZ *jets like a deflating balloon into outer space.)*

General Johnson: Just goes to show you, even the biggest threat to Civilization As We Know It has an Achilles Heel.
Dr. Brother: Or toe.

With the world safe again, I wanted to test our symbol of Truth, Beauty and the American Way for further injuries. There was only one way to determine the depth of her internal injuries, and I was determined to help treat her. I cushioned her naked body with the garments, raised my probe

---

(The PRODUCER *from* Good Morning, Audience *stands over them, younger, trimmer and with a full head of black hair.)*

Producer: I like your style. You have a Big future ahead of you.
Thorpe: I'm just trying to do my civic duty, sir.
Producer: I wasn't talking to *you*, twerp.
Andale: *(giggles)* You mean, little old me?
Producer: Little Old Big-In-All-The-Right-Places You. You just have to pass the Screen Test.
Andale: *(giggles)* When do I take it?
Producer: Right here and now. *(Pushes* THORPE *out of the way.* ALEXANDRA COLDBREATH *steps forward, lifts the* PRODUCER *by the back of his collar.)*

Coldbreath: What about me, mister? I wasn't exactly a wallflower.
Producer: I have just the part for you. You can be the star of *Attack of the Fifty-Foot Woman.*
Coldbreath: Sorry. The role's not big enough for me. *(Drops him into the rubble of a network TV building.)*
Thorpe: Let's get out of here while we can. *(Grabs* PAMELA ANDALE *by the hand. Rolling credits blur their run into the sunset.)*

[] [] []

(Medium shot, the schlub.)

Hi. My name is Ralph Putz, and I'm a Role Model. Now, most of you are used to seeing celebrity Role Models, like I was up until a few hours ago. I'm here to tell you that just because you're a celebrity today doesn't mean you'll be a celebrity tomorrow. Fame and glory are fleeting things. But they aren't everything. Neither is money and neither is power. But if you put fame and glory and money and power together, then you have everything. And as your Role Model, I'm going to show you how you can have everything, or my name isn't Ralph Putz and I'm not a Role Model.

While the World Gently Sleeps

(Medium shot. The AUTHOR.)

Now that you the TV audience have seen *Commercial Fiction* in all its various forms I'm here to tell you that we at the Family Discount Shopping Network are offering copies of *Commercial Fiction* to the first ten thousand callers at an UNBELIEVABLE discount. What's more, I'm going to handle each and every call personally, by arrangement with the Family Discount Shopping Network staff. They told me when I bought this hour that 3:00 A.M. was Prime Shopping Time for books like mine and that I could count on *a very minimum* of twenty thousand sales of *Commercial Fiction* before my hour was up. So, pick up the phone and start calling in those orders.

A little slow, huh? Well, my marketing director told me to carry a few side-products for the non-book-buying audience. So tonight I also have a Special 20% Off Sale on Pamela Andale T-shirts. Even if you don't like *Commercial Fiction*, you've got to like Pamela Andale.

Here's our first caller now. Hello! I beg your pardon. You're Pamela Andale's attorney and you're prohibiting me from unauthorized sales of the products she endorses? I really *do* beg your pardon. It just so happens that Pamela Andale is a *fictional* creation. That's right, *I* created her. Without me, she wouldn't even exist, let alone have a seven-figure endorsement contract.

Oh, she exists *now*, you say? Well, the way sales are going here, I should have a commission because I *created* her. That's right. Without me, she wouldn't even exist, never mind sell...what did you say, Pre-Wet T-shirts with *Breast Imprints* on sale at 1-888-555-5555? *What! No commission!*

Well, that takes care of that.

I still have T-shirts featuring Bubbles LaFlamme, Melissa the Model and Vanessa the Third World Gun-Runner in a special three-for-one sale. And here's our first caller.

What! The Women's Anti-Defamation League!

So much for that one. I guess I'd better put away the Alexandra Coldbreath Hair Shirt.

If I can't sell those, my sideline items are limited. How about a T-shirt of the former Host of *Good Morning, Audience?* It's got the classic shot you've seen on every newscast today.

What *is* this "under contract to Living Dead Productions" crap? Oh, you're also Necrophiliacs Anonymous? You sound like a real live one to me.

Looks like it's back to selling books. You really *should* read it. I'll even offer the first thousand copies of *Commercial Fiction* free, no strings attached. All you have to do is call.

NO! I DON'T HAVE ANY RALPH PUTZ T-SHIRTS, GODDAMMIT!

[] [] []

(Studio. Set of the All Night News Network. AUTHOR *is the Anchor.)*

Good morning. While the world gently sleeps, former media and business celebrity Ralph Putz is working to regain his former status. Putz, who late last evening learned he lacked the legal authority to purchase the United States government, has more than compensated for his losses. He has renewed Russia's commitment to the Cold War, strategically placing missiles throughout northern Siberia for a strike over the North Pole. He has armed Third World countries in South America and Africa. Informed sources say that he has also purchased Venus, Mars and Jupiter, and that negotiations for purchasing the Milky Way galaxy are taking place as we speak. For Putz, who started his career as a Role Model for underprivileged youth, this is proof positive that You Can't Keep a Mediocre Man Down. Oh, we're being cancelled? Good night.

A Life in the Day of (7)

I heard a fly buzz when I lived.

6:44 a.m. zzzzzzzzzzzzzzzzzzzzzzzzzzzzzzzzzzzzzzz

spooned around her Florida motel neon flickering outside window bodies naked on an empty beach sunset moon felt her softness what a way to

6:45 a.m. ***BZZZ-BZZZ-BZZZ-BZZZ—BZZZ-BZZZ-BZZZ-BZZZ***
BZZZ-BZZZ-BZZZ-BZZZ—BZZZ-BZZZ-BZZZ-BZZZ

WHAP!

"FUUUUUUUUUCK!"

The Real Way to start the day: kill the fly buzzing its Tuesday reminder. Waking cry of existential dread. Bounce out of bed. The pretense of vigor quickly vanishes as the dizzying whirl from reclining to standing settles like dead weight reminding me I'm not fully awake no matter what my position. The rush toward the coffee maker remains a trudge, the attempt at speed forced but ineffective. Nausea: not so much at being alive as at my options. Do I:

(1) Say fuck it & go back to sleep?
(2) Call in well, stay home & write?
(3) Go to work, feel sick & write?
(4) All of the above?
(5) Some of the above in various permutations?
(6) None of the above?

The answer, as anybody knows, is

COMMERCIAL FICTION

is brought to you by

THE AUTHOR

(Medium shot. A SUBURBAN MOTHER *and her* SON *stare at the present page, hovering on display above a table in a Megabook$ chain store.)*

Son: Mommy, what's that?
Mother: That's the author.
Son: I want to see him. *(Steps forward, tears away the page, revealing a button on the left lapel of an Armani suit. The camera zooms to:)*

Showing Up
is
100% of Life